Soils of the
Desert Southwest

Lucy K. Bradley
Urban Horticulture Agent
U of A Cooperative Extension
4341 E. Broadway Road
Phoenix, AZ 85040-8807

SOILS

of the
Desert
Southwest

WALLACE H. FULLER

Author of *Management of Southwestern Desert Soils*

THE UNIVERSITY OF ARIZONA PRESS
Tucson, Arizona

About the Author . . .

WALLACE H. FULLER has authored numerous scientific articles in national and international journals and is co-author of, or contributor to, several books. He frequently has served as consultant and advisor in biochemistry and in the management of water and soils. A Fellow in the American Association for the Advancement of Science, the Soil Science Society of America, and the American Society of Agronomy, he won the E. B. Knight Award of the National Association of Colleges and Teachers in Agriculture for the best-written article in 1972. A biochemist and professor of soil science, he joined the University of Arizona faculty in 1948, later to serve as department head of Agricultural Chemistry and Soils for sixteen years. Previously he was with the Agricultural Research Service and the Soil Conservation Service of the U. S. Department of Agriculture, with Washington State University, and with Iowa State University. Fuller, a native of Old Hamilton, Alaska, received his B.S. and M.S. degrees from Washington State University and his Ph.D. from Iowa State University.

THE UNIVERSITY OF ARIZONA PRESS

I.S.B.N.-0-8165-0441-5
L.C. No. 74-79390

To My Mother

Contents

Illustrations

What the Subsurface Looks Like

Soil Classification

Appendix A. Chemical Cycles and Soil Composition

Tables

A Word From the Author

A thin rind of loose material covering the continents of the earth is all that stands between life and lifelessness. Grit and grime, crumbling rock and decaying organic residue — abrading by wind and water — weather into soil — Mother Earth. This soft and yielding earth lives and continually changes under the forces of climate, having formed through the ages as a result of meteorological, geological, and biological action on rock. The soil not only lives, but it continually renews life as well. Animal and plant residues decay into simpler constituents, and nutrient elements again are made available for new life in a perpetual cycle.

Soils are not all alike, but differ, just as plants and animals do. They have easily recognizable characteristics which classify them readily into distinct bodies of nature. They acquire their individual properties from the forces which act upon them. Thus, desert soils differ from those of other climatic zones. This book describes the characteristics of desert soils of the southwestern United States and contrasts their characteristics with those of soils of more humid climates.

The increasing flow of people toward areas of arid and semiarid climates, and the rapid development of these lands for food production, home sites, and metropolitan uses, give reason for describing the earth from which desert life springs. Exotic fruits and off-season vegetables are a part of almost every household. Desert soil produces these products. Homeowners and residents who move from colder climates search for help in the establishment of a living foothold in a new and fragile environment. This environment is delicately in balance with an alternating benign and harsh climate: comfortable winters and hot summers; crystal-clear air and dust storms; cool valleys and sun-baked playas; arroyos, now dry, now storm-choked with booming boulders and mud flows; and flower-carpeted mesas dried to mirage-edged sand wastes.

To successfully grow plants, man must know comparatively more about arid climate soils than he does about humid climate soils. He should be aware of an ecology which can be readily altered or eroded by the invasion of the plow, overgrazing by domestic animals, and bulldozing by the "development" speculator. The husbandry of water resources is a necessity in the desert. New management practices must be mastered to prevent soil from going out of production from salt and/or alkali accumulation, chlorosis, crusting, compacting, or droughting — to name only a few problems which arise from mismanagement. Soil and water management practices necessary to maintain continued plant growth are complicated and not easily understood.

The need for sound soil and water management practices to perpetuate acceptable plant production also is as necessary for the homeowner with only a backyard to manage as for the commercial crop grower. The principles and practices are fundamentally the same. A knowledge of the soils, their characteristics, and the way they behave to the host of man's treatments is necessary for acceptable living in the desert.

Before soils can be made to respond favorably, a knowledge of their characteristics — biological, chemical, and physical — as well as of their surface and subsurface morphological nature must be acquired. This book describes typical soils of the hot deserts of the Southwest and draws attention to the specific characteristics which distinguish them from or relate them to humid soils. There is no attempt to bring together the many facets of the complicated science of soil and water management, which is an immense story of its own.

I am indebted to my Soil Science colleagues who gave cheerfully of their time and expertise in reading and making suggestions in the development of this book.

The University of Arizona provided invaluable facilities and other support through the Agricultural Experiment Station. The critical reviews by Thomas C. Cooper of the Journalism Department and by David M. Hendricks, Harm H. Havens, Arthur W. Warrick, and Kenneth K. Barnes of the Department of Soils, Water, and Engineering at the University of Arizona were very helpful. Encouragement by Kenneth K. Barnes, Head of the Department, during the writing of the manuscript was most heartening.

1. A Selected Desert

Within the general area of the southwestern United States are many individual large and small desert areas, surrounded by semiarid grasslands and, at higher elevations, by subhumid forests. Margins of the deserts cannot be clearly defined, because they expand during a series of dry years and contract in wet periods. They fall within the general boundary of the double lines on the accompanying map (Fig. 1.1). The boundaries are moving, transitional zones and alternate between semiarid grassland and arid desert following periodic shifts in climate. For convenience, geographers have classified the Southwest desert as consisting of two large land areas, called the Sonoran and Chihuahuan deserts, based primarily on differences in vegetation.

The Yuma subdesert, which represents the heartland of the extremely dry and hot Sonoran Desert, has a mean annual rainfall of about three inches. It often is referred to as a "true" desert. The Yuma subdesert is selected for detailed description in this chapter to represent what is meant by a true desert. To describe all of the subdeserts in the Southwest in their many detailed variations would serve only to burden the text. Moreover, soils of this selected area are broadly representative of many of those in other desert valley floors and on terraces and mesas along rivers and streams in all of the arid Southwest. Figure 1.2 depicts topographic features associated with desert lands and referred to in this volume.

PHYSICAL DESCRIPTION

The Yuma subdesert is essentially a comparatively smooth plain, interrupted by low, narrow mountain ridges of bare rock. Its lowlands are broad, fairly flat belts of sandy soil, with alluvial fans of rock debris nearer the mountains. Along the Colorado and Gila rivers, somewhat flat, narrow strips of valley soil support luxurious

[1]

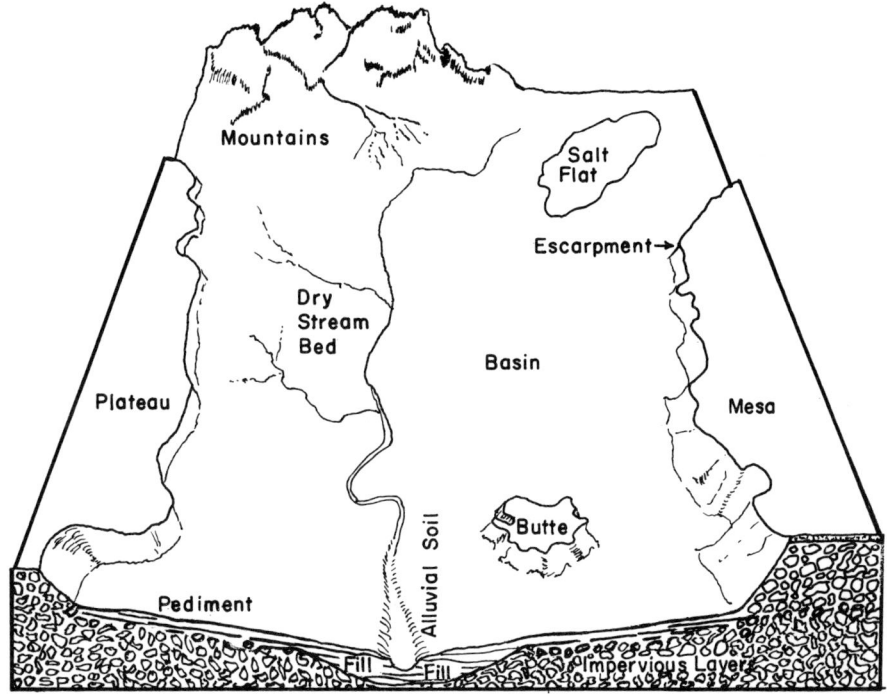

Fig. 1.2. Topographical features of arid lands.

plant growth where water is available. Here the surface is smooth, but broken in places by sand dunes and old river channels. The elevation is from 90 to 200 feet above sea level. The lowest levels are along the Colorado River and at its confluence with the Gila.

Away from the valley floors, the Yuma subdesert rises 40 to 75 feet to a broad, smooth, sandy terrace. That adjacent to the city of Yuma is called the "Yuma mesa." The terrace is separated from the valley floors by a well-defined bluff. The elevation transition between valley and terrace is more gradual along the Gila River than along the Colorado.

Beyond the terrace and extending mostly north and south is the Yuma plain. It is interrupted by many very rugged but comparatively low and narrow mountain ridges having a northwest-southeast trend, separated by broad lowland belts interspersed with a few small, partly buried hills, or buttes, all of basaltic nature. It is of much greater extent than the Yuma mesa.

Soils are marked by geologic history. The nature of the geologic debris, rock and stone, from which soils finally form, along with other factors such as climate, vegetation, micro-organisms, and time, give a soil its individualistic character. The geologic history that left the parent material from which the present soils originate does not lack for variety. The Yuma area has been exposed to alternating periods of humid and arid climate, and invaded by seas, lakes, rivers, and desert wasteland dunes — a fascinating story.

The soils of the terrace and plain are sandy, having been formed, in large part, by wind-blown particles from the river bottoms. They are called *aeolian,* since they are wind-laid. The valley lands, on the other hand, are laid down directly by river waters and are called *alluvial.* Near the mountains, broad, alluvial fans (water-laid deposits) of geologic debris also accumulate in various degrees of particle size (rocks, stones, gravels, sands, silt), depending on the volume and rate of flow of water in the outwashes.

VEGETATION

Only a sparse growth thrives on the terrace and plain. Creosote bush (often erroneously called greasewood) is most prevalent. Other plants include bursage (false ragweed), bunchgrass (tobosa), and jointfir (a bush also called Mormon tea). The arroyos, or washes, are scattered with palo verde, mesquite, and ironwood. Saguaro and cholla cacti and ocotillo grow on the mountains and slopes approaching the mountains. Uncultivated valley soils support cottonwood, willow, arrowweed, salt bush, seepweed, pickleweed, desert sage, salt cedar, and some creosote bush. Figures 1.3 and 1.4 show many

Fig. 1.3. Footslope and low ridges showing arid-land vegetation pattern in the Southwest.

Fig. 1.4. Typical vegetation in the Sonoran Desert.

of the plants in two typical desert landscapes. With the damming of the rivers upstream, some of the more moisture-demanding weeds and trees along the Gila have almost disappeared.

SOILS

The soils of the Yuma subdesert, for the most part, have developed on either wind- or water-transported material. As is usual under strictly arid conditions, the weathering is more a physical than a chemical process. Physical weathering includes fracturing by expansion and contraction as a result of heating and cooling and wind and sand abrasion. These processes are slower than the chemical weathering, freezing, and thawing action in cool, humid climates. Soils form poorly or not at all on solid rock outcrops, in contrast to what is commonly observed in humid climates.

Not all of the Yuma subdesert soils have reached their final state under the present climate, but are continually undergoing change, even though such change is extremely slow. However, because of irrigation practices, more rapid soil changes are occurring. For example, soils in home lawns accumulate organic matter and salts, and often are more compact than when in the virgin condition. Soils respond to their environment and are a product of their environment. What are the soil-forming factors responsible for desert soil characteristics, and how do they interact in soil formation in the desert? Since population centers in the United States are predominantly in humid areas, people have a greater familiarity with humid-area soils than they have with desert-area soils. Thus, it is important to make comparisons between the two as a matter of reference.

The most important factors that function in the transformation of rock into soil, in its broadest sense, are (a) climate, (b) vegetation, (c) topography, (d) parent material, and (e) time, plus accessory factors such as might arise from a high water table and seasonal variations in climate, and in man's treatment. When rain is limited, these factors act very slowly. Chemical dissolving of geologic minerals and moving of substances in and out of the soil are almost at a standstill under unirrigated conditions. Formation and reformation of clays are slow. Clay occurrence is primarily a result of wind transport of fine dust away from a humid climatic area, or clay formation during a humid cycle in the history of an arid area. The sand of the Yuma mesa is very low in clay, and chemical weathering that results in the formation of clays is not perceptible. Expansion and contraction from temperature changes result in the slow, mechanical breaking of rock, gravel, and sand into further smaller particles. The infrequent but heavy rains speed up the chemical actions of solution, carbonization, and oxidation, but only for brief periods.

Climate not only affects the speed of chemical reaction, but it controls the density and kinds of plants, animals, and micro-organisms. The sparse plant cover of the desert provides a very small return of organic residue to the soil, and, as a consequence, the accumulation of organic matter is small.

Desert soils rank among the youngest of soils, i.e., they are in their early stages of development, except where past history includes a humid climatic cycle. Except where recent wind-blown or water-laid materials exist in significant depths, such as in the Yuma sub-desert, soils in the Southwest are dominated by humid weathering which occurred during earlier climatic eras. Consequently they resemble, in their morphologic and genetic characteristics, humid soils more than they do arid soils. The more recently imposed aridity complicates characterization of desert soils.

The designation of "young" or "poorly developed" soil is a relative term. As used here it refers to soils formed under recent arid conditions and not yet fully developed. The Yuma sand (Superstition sand) is a good example. Near the mountain ridges the sand is thin and overlays water-deposited land or remnants of soils developed under humid climatic periods (Fig. 1.5). These soils have varying characteristics, stratifications, and mineral composition. Layers of sand, silt, and clay of varying thickness stratify throughout the root zone and compose much of the subsoil. Water-deposited layers of caliche (cuh-leech-ee) or lime and/or stone and gravel also appear under these thin sandy soils. Figure 1.5 illustrates the usual occurrence of wind-deposited sand, which is deep near the river bed and shallow on the footslopes and slopes of the ridges of parent rock.

MAN'S INVASION*

Indian tribes inhabited the Yuma area for possibly several thousand years before the first Spanish *entrada*. Their population density and movement followed both the Colorado and Gila rivers. The city of Yuma is located at the confluence of these two major streams of the Southwest. Remnants of Indian irrigation systems may be found even today. The porous sand of the terrace and plain was not cropped by these early people. The Hohokam Indians of central and southern Arizona developed an elaborate system of controlled irrigation along the Gila River as early as 800 A.D. Thus in the early human history of the Southwestern desert, the growing of food and possibly fiber crops became a part of the social economy. Compared with today's sophisticated irrigation techniques and crop production, the early aborigines' agriculture was crude and only partially effective in sustaining population development. By the time of the *conquistador* invasions, the early elaborate irrigation system had deteriorated to almost nothing. Some evidence indicates that mismanagement of soils and water, with the consequent deterioration associated with

---

*For a detailed early history discussion of the people inhabiting the Sonoran Desert see Roger Dunbier, *The Sonoran Desert: Its Geography, Economy, and People.* University of Arizona Press, Tucson, 1968.

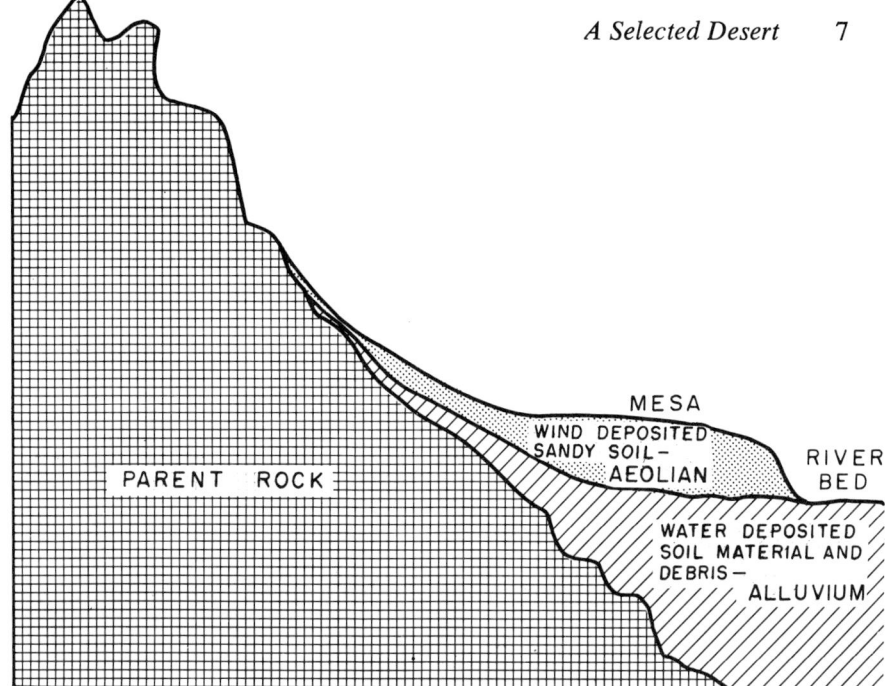

Fig. 1.5. Diagrammatic scheme of the relationship between topography of the land and the original parent rock, the air- and water-deposited materials, and the young soils developing from these deposits.

salt relationships and failing crop production, may have been partly responsible for the abandonment of irrigated land. Sauer and Brand* estimate the population of Indians, prior to the Spanish settlement in 1521, as being small, probably one or less per square mile. Of the many nomadic tribes and those few agriculturally oriented tribes that entered the Yuma subdesert, not many remained as settlers.

Travel and trade routes developed across the Yuma desert partly because the Colorado and Gila rivers join there, and partly because the Gulf of California, a short distance to the south, permitted coastal maneuvering. The early Spanish penetration into the Sonoran Desert and the Yuma subdesert was marked by the conquest of local tribes and their lands.

More permanent settlement and Spanish conquest of the desert took place during the Mission Period, i.e., in the early 18th century. Unlike the failing *ecomienda* system of the military force, the mission system set a pattern for occupation and settlement through peaceful

*Carl O. Sauer and Donald Brand, *Prehistoric Settlements of Sonora.* University of California Press, Berkeley, 1931.

means. The mission settlements became the major centers of population, irrigated agriculture, and community life.

The pattern of development of towns or villages in arid lands was influenced by the quality of soil. Early Spanish settlements generally consisted of a grant, usually four square leagues of land in the form of a rectangle with the first point being designated as the town square or plaza. Additional areas were laid out in the form of grids with the town square as the center. The church and major public buildings were located around and facing the plaza. This ideal plan was not always followed because of adverse topographical and soil conditions associated with both agriculture and mining settlements. Usually the village was planned laterally, paralleling streamways where soils are more favorable.

Father Eusebio Kino, explorer, who died in 1711, distributed wheat, cattle, and horses to Indian tribes along the Gila and Colorado rivers, but he never developed fully those areas as he did the borderlands of the Pima Indians at Tucson and points south. Not until the gold rush in 1849 was the east-west route down the Gila to the Yuma crossing of the Colorado traveled significantly.

Anglo-American settlement of the Sonoran Desert in Arizona took place more than 300 years after the first Spanish establishments. Mining attracted the Anglo-Americans more than agriculture. The full impact of man on the soils in the Yuma subdesert was not felt until the early 20th century. Small parcels of irrigated land along both the Colorado and lower Gila were established and abandoned sporadically until the late 19th century. The building of the Alamo Canal began in 1902. This delivered water to Imperial Valley, California, via natural drainageway, partly in Mexico and partly in the United States. In 1905 the Colorado flooded into and overran the canal and drainageway, filling the Salton Sea depression to a maximum depth of 72 feet and covering 330,000 acres before it was finally diverted back to its delta. Each year the Colorado River deposited many acre-feet of sediment to the delta. There was constant fear the Colorado River would overflow and inundate the Imperial, Yuma, and Mexican desert areas, located near or below sea level, and drown its populations. The Hoover Dam Project, which impounds water in Lake Mead and provides hydroelectric power as well as streamflow control, was authorized in 1928 and completed in 1935. Further down-river, the sediment-laden water

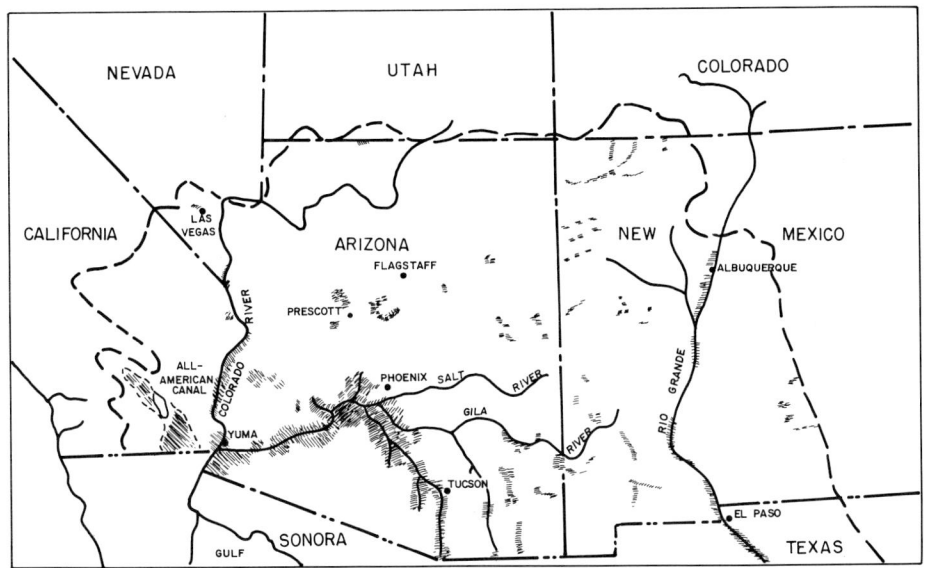

Fig. 1.6. Irrigated lands in the desert Southwest. Hatched areas define the extent of irrigation.

required the building of a desilting works in order for it to be used for irrigation. Imperial Dam, at the intake of the All-American Canal that delivers water to clayey soils of the Imperial Valley, was built for this purpose. On the other hand, sediment brought to the sandy Yuma mesa prior to building of the desilting works had a favorable influence on these porous soils. The water-holding capacity and chemical bonding of the fine silt made the sandy mesa more agriculturally productive. These and other dams have encouraged extensive agricultural land development in the Yuma subdesert. The environment has been considerably changed, from an area to be avoided to one of the fastest-growing population areas of the United States. Irrigation agriculture flourishes (Fig. 1.6).

Indeed, water has made the desert Southwest as a whole flourish, just as it has the Yuma subdesert. Raw desert lands from the Pecos to the Imperial Valley explode into cities so rapidly census figures become highly distorted before they can be published. Soil resources not only influence the way of life of the desert Southwest but are involved with the details of living of each individual. Thus with a general knowledge of land and soil conditions in the heartland "true" desert — Yuma subdesert — as a background, the remaining part of this book will describe in some detail soil and land conditions of the desert Southwest as a whole.

2. How Soils Differ

All the green thumbs in gardening cannot change the differences that exist among soils, which are as great as those among trees or birds or insects. Soil composition varies within a single climatic region almost as much as among different climatic regions. Wide differences exist in fertility and productivity. Even when brought to the same level of fertility by adding nutrients, different soils do not necessarily produce identically. Soil variations are inherited during formation from differing geologic parent materials and/or from the differing conditions under which soils develop. Soils vary in depth of the surface organic layer and total depth to parent material. Yet some soils of the desert Southwest are as fertile in the subsoil as they are in the surface layers. Unlike the more developed humid-climate soils, removing the topsoil of desert valleys by cutting, filling, and leveling does not necessarily impair its productivity and may improve it. On the other hand, where soils are shallow because of underlying caliche, rock, gravel, sand, or hardpans, stripping the surface exposes a poor medium which always creates a problem for growing plants.

A knowledge of soil and water behavior is essential for successful gardening in an arid climate. Although it is well known that a scarcity of water will cause salts to accumulate in soils, successful gardeners also have learned that plants can be damaged by excessive watering. Without provision for drainage, excessive water in most soils creates a water-logged condition in the subsoil which leaves plants stunted and yellow (chlorotic). When over-watering is continued a "perched" or "high" water table can develop. Salts collect on or near the surface, moving up by capillary action from the subsoil water. Such soils soon become so salty plants grow raggedly or fail completely to grow. On the other hand, for those who know soils and how they interact with water, many beautiful gardens and landscapes flourish.

MINERAL MATTER*

Kinds of minerals in soils

Mineral matter constitutes the bulk of soils in the desert Southwest. Mineral materials originate from rock. When rocks are reduced to small particle sizes in the extreme outer layer or crust of the earth, principally by weathering, they form into soils. Not all minerals change chemically by weathering. *Primary minerals* that remain relatively unaltered by weathering do so because of their chemical and structural nature. For example, precious stones like rubies, garnets, and zircons remain resistant to weathering, whereas the minerals in the original rock surrounding them dissolve and crack or wash out and the rock crumbles, freeing the resistant minerals and stones. "Sand garnets" may readily be found completely unchanged in sands collecting in stream beds and arroyos. Those found along Sabino Canyon near Tucson are a good example. *Secondary minerals,* called clays, form from the less resistant rock material. Thus soils are an organized body containing both primary and secondary minerals. The majority of the minerals in soils falls into the following technical groups:

1. *Oxides.* Dominated by quartz, which is prominent in sand. Iron oxides in various degrees of hydration give soil red to yellow colorations.
2. *Silicates.* The largest number of minerals in soils are silicates. They are in combination with numerous elements of nutritional significance to plants.
3. *Carbonates.* Calcite and dolomite are the predominant carbonate minerals in soils. They are often added to acid soils where they are low, but not to neutral or alkaline desert soils where lime accumulates naturally.
4. *Feldspars.* Feldspars are an important constituent of clay.
5. *Micas.* The glittering flakes of mica in rocks and soils can be readily identified. Some clays are micaceous in nature.
6. *Sulfides.* Pyrite fragments (often mistaken for gold) may be seen mixed in sand and gravel as well as silts.

*Mineral as used here in its broadest meaning refers to "any chemical element or compound occurring naturally as a product of inorganic processes." Webster's Dictionary, Ed. 19.

7. *Phosphates.* Apatite, a dark brown mineral, mixed widely in soils in small amounts, is the most prominent phosphate mineral.

Minerals do not weather at the same rate. Oxides and mica, for example, tend to accumulate in soil, as compared to the other minerals. Parent rock from which soils are derived contains different proportions of these minerals. Since the materials which make up the largest part of the soil originate from these minerals, it can be expected that soils may differ according to the proportionalities of their mineral components.

Primary minerals

Primary minerals predominate in the sand fraction of soils. On an equal weight basis, they are less reactive than the secondary clay minerals. *Quartz, feldspars,* and *hornblende* constitute the three most abundant primary minerals in the earth's crust. The proportion of these minerals in soils is related to the composition of the different igneous rocks as parent material.

Clay minerals

During the weathering of rocks and minerals, new minerals form. The most prominent are the *clay minerals.* Though these secondary particles are small, they dominate the chemical, physical, and nutritional behavior of soils. Clay minerals form so slowly, however, it is difficult to evaluate the effect of desert climate on the particular characteristics of clay. Desert soils of the Southwest derive their clays, for the most part, from earlier periods of humid climate. Moreover, the alluvial and wind-deposited desert soils have been transported from more humid areas or from formations along river channels. In any event, desert soils are greatly mixed, both in derivation of soil-formed clay and geological rock materials. This great mixture of materials is one of the most distinguishing characteristics of soils of the desert Southwest. Two soils of different clay minerals, and consequently different chemical and physical properties, may occur within 100 yards of each other.

Lime (Calcium carbonate)

One of the most prominent mineralogical and chemical differences between soils of humid and arid climates is the dominance of lime and other calcium compounds in arid climate soils. Under humid conditions lime is added to soils to obtain maximum plant growth by counteracting acidity. In desert soils lime appears in excess as veinings, nodules, and/or solid layers often called *"caliche"* (Fig. 2.1). Fortunately, a host of ornamentals, as well as vegetables, grow well in the presence of lime. In fact, The University of Arizona at Tucson is located on a caliche knoll. A great variety of ornamental trees, shrubs, and bedding plants cover the grounds, though some unusual cultural practices and care in proper irrigation are necessary. Plants requiring acid soils are not grown except in planters or small beds, where acidulating agents and certain trace elements such as iron can be supplied. Acidifying soils of the desert in the present 1974 economic circumstances, except in planters and small, pampered garden plots around the home, is not practical and may

Fig. 2.1. A shallow soil developed during a period of humid climate, later becoming part of an arid climatic area. Lime (caliche) is several feet thick.

do more harm than good to the soil. The difficulty of maintaining acid conditions is in itself discouraging, to say the least. The necessity to irrigate with water which leaves a residual alkalinity, and the poor adaptability to a hot, dry climate, dictate seriously against growing certain cool-humid climate plants, even if the soil can be made suitable for root growth. When a plant is moved from its natural environment to one in which it is not adapted, trouble begins. Special attention must be provided to make the soil more like the one in which the plant grows naturally. Fortunately, a large number of ornamental shrubs and urban landscaping plants, vegetables, and fruit and nut trees grow well in most desert climates, and excellent choices of adapted plants already exist.

The desert gardener is confronted with new and often puzzling soil conditions that are not a part of plant production in areas of higher rainfall. Desert soils usually contain excesses of certain constituents, such as lime and gypsum mixed with other salts, not common to other climates. In humid regions, soil problems are solved by adding something which is lacking to the soil. In arid regions this doesn't always work. More often than not, there are excesses of undesirable constituents and something must be *removed* to correct problems. Leaching (washing out salts) with water is the most effective practice to remove excess salts. Also, leaching of sodium soils by the addition of gypsum, the calcium of which will displace the harmful sodium element, often is required before preparing some soils for planting.

Calcium primarily, and sodium and potassium to a lesser extent, dominate chemical reactions in soils of arid lands. In soils of humid lands chemical reactions are dominated by hydrogen, aluminum, and iron. Therefore treatment of soils to develop or improve their productivity is entirely different in the two climatic areas.

Nitrogen

Soils differ considerably in natural fertility. Nitrogen varies more than any other single plant nutrient. Soil organic matter contains most of the reserve nitrogen which eventually becomes available to the plant, therefore if organic matter is low, so is nitrogen. Soils of the desert Southwest are notoriously low in organic matter. Small quantities of mineral nitrogen are made available by soil bacteria, which convert insoluble nitrogen gas of the atmosphere to soluble forms of nitrogen. Algae also convert atmospheric nitrogen into a form available to plants. Despite heat and drought, algae remain viable on the surface of the desert as a crust and become quickly active during periods of infrequent rains (see Fig. 2.2). A high proportion of desert plants belong to the legume family (palo verde, acacia, mimosa, ironwood, etc.). In legumes the root nodule bacteria (*Rhizobium*), through a symbiotic (cooperative living) relationship with the plant, combine atmospheric nitrogen with hydrogen which then becomes available for plant growth (this process is technically called nitrogen fixation). In the nitrogen cycle (Appendix, Fig. A.1) micro-organisms play the leading part in nitrogen transformations making mineral nitrogen available to plants.

Fig. 2.2. Desert algae encrusting quartz stone. The crust is formed on the side that is in contact with the soil. Light transmitted through the opaque quartz provides energy for photosynthesis. The algae belong to the blue-green group.

Application of nitrogen fertilizer is necessary to maintain the high supply demanded by vigorous plant growth. Since plants derive their nitrogen from inorganic compounds like ammonium and nitrate salts regardless of whether they come from mineralization of organic matter or from a commercial fertilizer bag, the cheapest source is the kind to apply. In this respect it is a lesson to note the experiments in continuous growing begun in 1843 at Rothamsted, England, which compared *source* and *level* of nitrogen fertilizer: (1) 14 tons of barnyard manure a year; (2) no manure but fertilized with chemical fertilizer equivalent to the nutrients supplied by the manure; and (3) no fertilizer of any kind, gave the following results after 120 years: (a) 35 bushels of wheat per year per acre; (b) 35 bushels of wheat per year per acre; and (c) 13 bushels of wheat per year per acre.

Thus in 120 years of fertilizing no differences in yield could be found between the land receiving fertilizer from *organic* and that receiving fertilizer from *inorganic* sources.

Phosphates

Calcium phosphate minerals are abundant in the soils of the desert Southwest, yet phosphorus frequently is deficient for best plant growth. The principal mechanism for phosphate tie-up (unavailability to plants) in acid soil operates through the exposed chemical bonds of iron and aluminum located on clay minerals. In alkaline soil, fixation* or tie-up is by chemical combination with calcium to form slowly water-soluble calcium phosphates. Acidification of desert soils, therefore, tends to make the native calcium phosphates more available (soluble) to plants. This should not be taken to imply that desert soils should be made acid. There are other effects of excess acidification which are not favorable.

Desert soils differ in quantity of organic as well as inorganic phosphorus. Organic phosphorus, like organic nitrogen, provides a pool of reserve phosphorus which is released by decaying for plant use. Organic phosphorus accumulates more rapidly under cool-wet than under hot-dry conditions. The amount of organic phosphorus is roughly proportional to the organic matter content of the soil (Fig. 2.3). A rough estimate of organic matter may be calculated from the

---

*Fixation is a term applied to plant nutrients which become unavailable to plants by their chemical or biological combination with other elements, forming compounds insoluble or slightly soluble in the soil solution. Not to be confused with nitrogen fixation mentioned above.

ORGANIC PHOSPHORUS–PPM

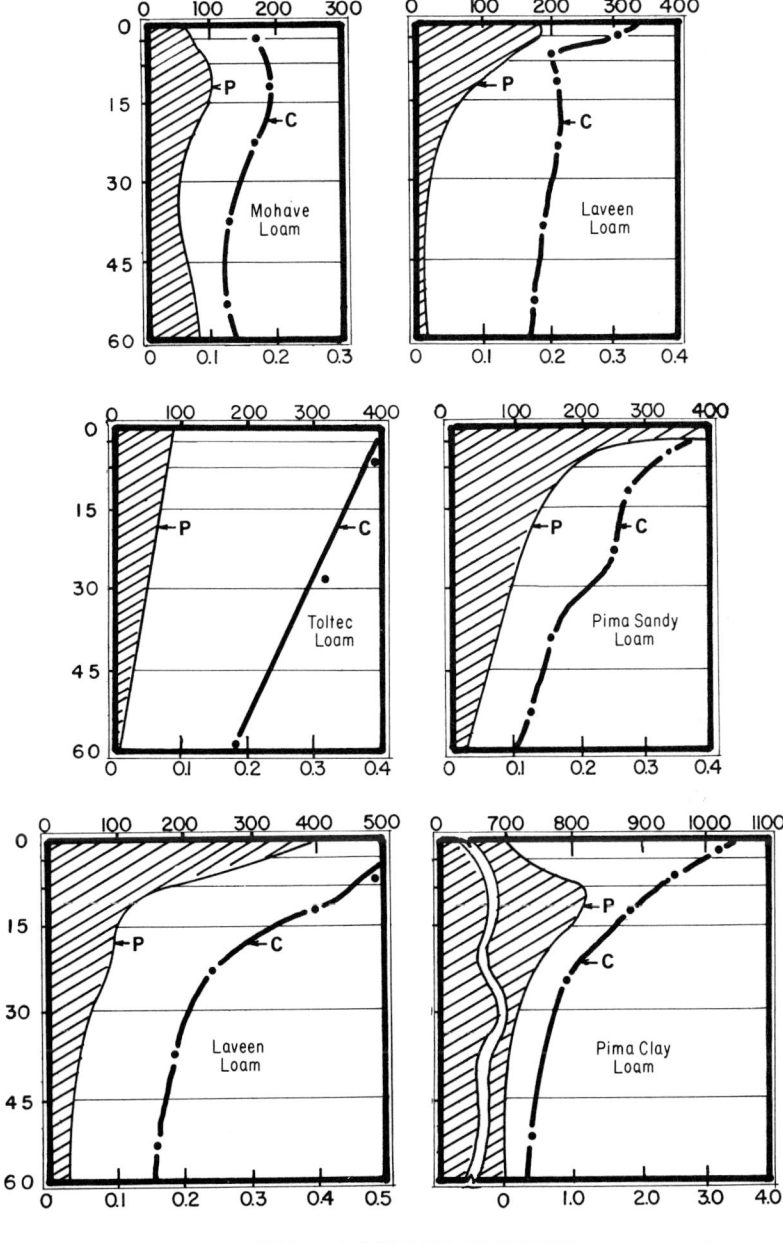

DEPTH IN CENTIMETERS

ORGANIC CARBON–PERCENT

Fig. 2.3. The organic phosphorus of some desert soils and its relationship to soil organic matter expressed as carbon/phosphorus. (From Tucker and Fuller in *Food, Fiber, and the Arid Lands,* McGinnies *et al.,* eds, University of Arizona Press, 1969)

organic carbon content in Figure 2.3 by multiplying the carbon content of soil by 1.7. Soils of wet-cool climates, despite their greater quantity of organic phosphorus, generally are more deficient in plant-available phosphorus than are desert soils. The inference is that the calcium phosphates of desert soils become more readily available for plant use than those of iron and aluminum in wet-cool soils, and that this more than makes up for differences in mineralization, or release of phosphorus from the organic matter. The difference in phosphorus availability represents a savings for gardeners in the desert Southwest, since they do not need to purchase as much commercial phosphate over the years as do gardeners in other climates. Furthermore, because of the slowly soluble nature of calcium phosphates and their resistance to leaching, fertilizer phosphate remains available to plants in the soil for several years after application and does not require annual replenishment. Further information on the phosphorus cycle appears in the Appendix, Figure A.2.

Potassium

Soils of the desert Southwest are well-supplied with potassium. Potash (a potassium fertilizer) fertilization of soils is not often required. Deficiencies have been found mostly in very sandy soils. The original parent material of most soils is high in mica minerals which are rich in potassium (Fig. 2.4). Only a few soils have excess potassium. Excess potassium has an adverse influence on soil structure similar to excess sodium. Soils containing excessively high potassium may be found in the Salt River Valley of Arizona in old stream channels such as those near Higley, although most of the potassic land now has been reclaimed by procedures similar to sodic soil reclamation, using gypsum with leaching.

Micronutrients

The desert soils of the Southwest are adequately supplied with the micronutrients *copper, molybdenum,* and *manganese* in available forms. Additions of *iron* are required for a few plants and a few soils. Deficiencies of iron occur most often in limy soils and where caliche is near the surface. Some home garden and landscape plants show chlorosis at certain times of the year and when watering is mismanaged.

Desert soils contain an abundance of *sulfur* and *boron*. Additional information on sulfur transformations and its mineralization

Fig. 2.4. Flakes of mica on a desert soil surface catch the sun. Because of the abundance of mica (vermiculite, potash-bearing minerals) naturally present in the geological material, potassium is not deficient for plants in soils of the desert Southwest.

appear in the sulfur cycle given in Appendix, Figure A.4. In desert soils excesses of sulfates and borates, particularly the calcium and magnesium combinations (magnesium sulfate is the well known Epsom salts home remedy) are more of a problem than are deficiencies.

Some crops, such as dry beans in the northern Arizona cold desert, become chlorotic from deficiencies of *zinc*. No other great area of zinc deficiency occurs in the desert Southwest, except occasionally in some narrow valleys, particularly in pecan or peach groves.

Total dissolved solids (Salinity)

The total dissolved solids represent, primarily, what are more commonly called *soluble salts*. Some of these are calcium sulfate, calcium chloride, calcium phosphate, magnesium sulfate and chloride, sodium sulfate and chloride, and a variety of bicarbonates and nitrates. Soluble salts accumulate in desert soils and leach out of those in humid climates. Some accumulated salts may be toxic, i.e., they adversely affect plant growth in small amounts. Other salts in larger concentrations may adversely retard plant growth through inhibition of osmotic water uptake. In unusually salty soils, the cells of roots physically collapse, injuring or destroying the cells. Certain other salts such as the calcium and magnesium carbonates and sulfates (lime and gypsum) form relatively insoluble residues which precipitate* at such a low level of solubility that they are not injurious to adapted plants. The danger of salt accumulation in the root zone of plants under arid conditions requires that a certain amount of irrigation water be used for leaching salts to below the root zone. In humid climates, leaching is accomplished naturally by the high rainfall.

Arid-land soils are often further complicated with salts from irrigation water. Water quality is an important factor to consider in growing plants in the desert. Well and surface water in the Southwest contain salts varying from a fraction of a ton to several tons per acre-foot depth. Rainwater is almost free of salts.

Perhaps the most serious salt problems occur when soils receive too little water. Where water is insufficient to keep dissolved salts washing downward they move upward, even to the surface, by capillary action. Accumulation of salts also can take place anywhere within the root feeding zone with little or no visible effects on the surface. Sometimes locating the salt concentration is so difficult that only chemical soil analyses can identify the problem area. Because of soil texture variability at different depths in valley soils, salt concentrates in a stratified condition. For example, clay layers have a tendency to accumulate a higher level of salt than sandy layers. Within the root zone of a plant, several layers of soil having different salt concentrations which exceed the tolerance level of the plant may occur. Two actual examples of salt stratification and texture interaction are provided in Table 2.1. Soil A is poorly leached. Salts have accumulated to the extent that most nontolerant plants die, and seeds germinate poorly or not at all. The fine-textured layers (clay) concentrated salts more than coarse-textured layers (sand). Salt accumulated on the surface.

---

*Precipitate is a word best understood by comparing it to the process of boiler scale formation and accumulation inside boilers, hot-water heaters, or teakettles.

TABLE 2.1

Effect of textural stratification and leaching on salt accumulation in two arid land alluvial soils.

Soil texture	Soil depth (in.)	Salts dissolved in saturated soil solution (ppm)
Soil A (poorly leached)		
Clay loam	0–8	11,477
Sand	8–24	7,638
Sandy loam	24–43	8,129
Clay loam	43–62	11,227
Clay	62–78	14,678
Soil B (leaching practiced)		
Fine sandy loam	0–14	1,273
Sandy loam	14–24	2,419
Clay	24–36	2,914
Sandy clay	36–78	3,658

Soil B is equally stratified with sand and clay. Adequate leaching with irrigation water moved the salts out of the root zone, prevented salt accumulation on the surface, and overcame the tendency of textural stratifications to differentially accumulate salt. Plants grow satisfactorily in soil B.

Boron and bicarbonates

Desert soils differ from those of humid climates in their accumulation in toxic concentrations of such salts as boron and bicarbonates. Fortunately, only a few areas in the Southwest contain these salts at plant toxic levels. When present, they require the adoption of reclamation practices and development of procedures for evaluating permissible levels in the soil solution for various plants.

Sodium (Alkalinity)

Sodium salts are so soluble they do not accumulate in harmful quantities in soils of humid climates nor in arid land soils except in limited areas. Some accumulations do occur as a result of man's poor soil and water management practices, others are found in uncultivated low spots where salt and water accumulate as runoff or in poorly drained hollows underlain by an impervious layer of clay or other duripans. Riverbottom clay and very fine sands often are high in sodium. Such soils are not recommended for landscape or garden purposes since they disperse on wetting, infiltrate water slowly, and inhibit root growth. Sodic soils are readily detected by measuring the pH value of the water-saturated soil paste. Characteristically the pH values of sodic soils range from 8.4 and up. Reclamation of these soils requires leaching of sodium below the root zone, and often the addition of gypsum as a source of calcium to replace the sodium. Some clay pan soils need drainage structures installed before leaching can be accomplished.

The term *alkaline soil* is used loosely to refer to soils having a pH reaction on the alkaline or high side of neutral pH 7.0, just as acid soils refer to those on the acid or low side of neutral. Plants are not adversely affected by alkalinity of the soil until the pH value exceeds about 8.4. Plants grow quite well between the pH range of about 5.0 to 8.4. Plants vary in adaptation to grow optimally in acid or alkaline soil reactions. Blueberries prefer an acid soil, whereas figs prefer an alkaline soil and lettuce grows well over a wide range.

Sulfur application counteracts excessive alkalinity in desert lands, just as ground limestone counteracts excessive acidity in humid areas. The adverse effect of excessive alkalinity also has been corrected by use of sulfuric acid, iron sulfate, calcium polysulfides, and manures. Prolonged use of acids or acid-producing chemicals will create an undesirable habitat for plant growth. Acidification should be practiced only on recommendation of experts knowledgeable in soil science and fertilizer practices of arid lands.

ORGANIC MATTER

Soils of the desert Southwest contain small amounts of organic matter compared with those of humid climates. The sparse vegetation and year round desert temperatures favorable for rapid decomposition do not allow the accumulation of organic matter in appreciable amounts. Soils in the Sonoran Desert vary, for example, from 0.1 to 1.0 percent organic matter as compared with 3.0 to 5.0 percent in humid soils of the Midwest. Even in small amounts, though, organic matter exerts a profound influence on the physical and biological characteristics of soil. The maintenance of this small amount is essential for maximum plant growth and most successful landscaping. Such practical factors as soil structure and the consequent water behavior, plant nutrient supply, and microbial activity are involved.

Continued soil productivity depends to a practical extent upon the replenishment and maintenance of some organic matter. The maintenance of a given, fixed level often is not practical, nor is it advised. The practice of incorporating readily available plant residues, manures, straw, leaves, composts, etc. into the soil is more important than the establishment of a fixed level. In irrigated soils, turfs, home gardens, and cultivated land, the organic matter level may increase substantially above that under native conditions.

Composition

Soil organic matter consists of partially rotted residues of the plants that occupied the land under native or cultivated conditions, the materials synthesized by micro-organisms, and the dead tissues and cells of micro-organisms. A part consists of living micro-organisms. Organic matter forms when plant materials, worms, insects, and debris of all natures decompose or rot. Decomposition is a desirable and necessary process for the recycling of plant food elements. Mineral elements, such as nitrogen, phosphorus, potassium, and micronutrients, are released. Carbonaceous substances degrade to carbon dioxide and water.

Five classes of compounds form the predominant part of soil organic matter. These are:
1. Lignin-like compounds (the less readily decomposable plant constituents).
2. Carbohydrate-related compounds (like microbial gums and slimes).
3. Nitrogenous compounds (derived from microbial protein and other less readily decomposable nitrogenous compounds).
4. Organic phosphorus compounds, part of which originate from microbial cells, tissues, and debris.
5. Growth regulators (growth-inhibiting antibiotics and growth-promoting substances, in trace amounts).

Though Southwestern desert soils, like other desert soils, are notoriously low in soil organic matter, the year round growing climate favors an abundant production of plant material where water is applied. An amazing amount of crop residue (2 to 20 tons/acre) decomposes each year in most irrigated agriculture soil. In fact, as much or more residue is recycled each year in an acre of irrigated arid-land soil as in an acre of humid-land soil. It is this recycling that gives life and productivity to desert soils.

Lignin-like compounds. Lignin is one of the most resistant of the major plant constituents. Although it undergoes rapid alteration by micro-organisms when first incorporated in soil, a portion remains that resists further degradation. It loses its identity as lignin during degradation and cannot be distinguished as such, though many of the original chemical properties remain intact. These fragmented products contribute substantially to soil *humus.*

Carbohydrate-related compounds. Compounds related to carbohydrates such as bacterial gums, slimes, and fungal products occur in soil organic matter. Microbial gums and slimes contribute to favorable soil structure formation. In this respect, the activity of soil micro-organisms has been related to formation of water-stable aggregates or crumb structure. Good soil structure in turn provides favorable water relations and aeration in the soil.

Proteins and related nitrogen compounds. Soil proteins (a major source of nitrogen in organic matter) come chiefly from microbial tissues. Soil proteins also are greatly modified from their original structures. The nitrogen components in soil organic matter are thought to be these modified protein compounds and complexes related to protein. Certain nonprotein-like organic nitrogen compounds have been identified in small quantities, also.

Organic phosphates. Soil organic matter also contains organic phosphate compounds. Desert soils of Arizona can contain as much as one-third of their total phosphorus in this form, although one-tenth is more common. The close relationship between soil organic matter and organic phosphorus is shown in Figure 2.3. Organic phosphorus compounds appear to be derived, in part at least, from soil micro-organisms. Movement of phosphorus to lower depths in soils is slow and takes place primarily in the organic (probably as chelates and microbial debris) rather than the inorganic form.

Growth regulators. Growth regulators are produced by soil micro-organisms in trace quantities. They are (a) *"cell-dissolving,"* as bacteriophages of the beneficial legume organism, (b) *growth-inhibiting antibiotics,* as penicillin, patulin, streptomycin, and terramycin, or (c) *growth-promoting,* as thiamine and biotin. Such substances produced by the native microflora exercise control over the ecology. However, the addition of micro-organisms *not* native to the soil is ineffective and they soon die out.

Functions

Most people know soil organic matter more by its functions than by its composition. Some of the more outstanding functions are as follows:

1. Direct contribution of nutrients for plants, such as nitrogen, phosphorus, and minor elements.
2. Indirect contribution to plant nutrition by making soil elements more available.
3. Food and nutrient supply for micro-organisms.
4. Soil structure development, and thereby water behavior and conservation.
5. Storehouse of elements attached to clay particles.
6. Biotic controls in soil through the presence of particular compounds now called "growth regulators" such as antibiotics and growth-promoting factors.
7. Effect on soil temperatures.

Plant nutrients. Organic matter contains such plant nutrients as nitrogen, phosphorus, and potassium, and trace elements of iron, manganese, copper, and boron. In fact, plant residues could be classed as near-perfect fertilizer. They need only to be decomposed to become available to plants. Plant nutrients in crop residues and soil organic matter are largely in organic combination and are mineralized or released slowly during the decay process.

In the absence of chemical fertilizer, soil nitrogen is almost wholly in the organic form. In unfertilized soils it represents the sole natural source for plant production. A rather constant ratio of carbon to nitrogen exists in all soil organic matter. In desert soils of the Southwest the ratio ranges from 8:1 to 10:1. This is narrower than in soils of the prairie belt of the Midwest, indicating the organic matter is more highly degraded or decomposed in the warmer climate. The importance of organic matter becomes more apparent when we realize that the nitrogen status is directly proportional to the amount of organic matter present. Thus, the higher the organic matter, the greater is the natural fertility status, and, conversely, the lower the organic matter, the poorer is the fertility.

Only small quantities of mineral nitrogen (nitrates and ammonium) may be found in the soil at any one time, and certainly not sufficient to meet plant needs. For example, the total nitrogen in desert soils may range between 400 to 1,600 lbs. per acre, though seldom are there more than a few pounds per acre of mineral nitrogen present at one time. Plants are dependent upon the conversion of organic nitrogen to inorganic nitrogen by soil micro-organisms, since plants cannot directly assimilate the nitrogen of organic matter.

The interrelationships between phosphorus and carbon and transformations of organic phosphorus into inorganic forms are similar to those for nitrogen, although the ratio of organic carbon to organic phosphorus (C/P ratio) is wider than the C/N ratio, varying from 19 to 38 (see Fig. 2.3). Most of the phosphorus of crop residues and soil organic matter is made available to plants slowly by microbial decomposition.

Food source for soil organisms. Soil micro-organisms, like other biological systems, require food to live. Organic matter supplies the necessary food for most of the soil micro-organisms, except for a few that use carbon dioxide of the air and inorganic nutrients. Plant nutrients are released as a result of the decomposition action by soil organisms. The organisms in turn obtain their necessary energy source from the degradation of the organic materials and residues that enter the soil (Fig. 2.5). Plants would starve for nitrogen and certain other elements if it were not for the mineralization action of the soil organisms.

The size and activity of the microbial population is determined by the amount of organic matter in the soil. Desert soils are not sterile; in fact, large microbial populations are found in all soils. Although soils low in organic matter may have small, active populations, as soon as organic residues are available and moisture is adequate, populations rapidly increase to very high values.

Life on this earth would not be possible without the soil micro-organisms to cycle and recycle available inorganic elements via the decomposition route. Without decomposition, plant residues would accumulate to such an extent that life as we know it could not exist (Appendix, Fig. A.3.).

Organic matter and plant residues act also as food for the larger soil fauna. Earthworms are considered to benefit the soil by their burrowing, soil mixing, channel action, and castings. Collectively, these

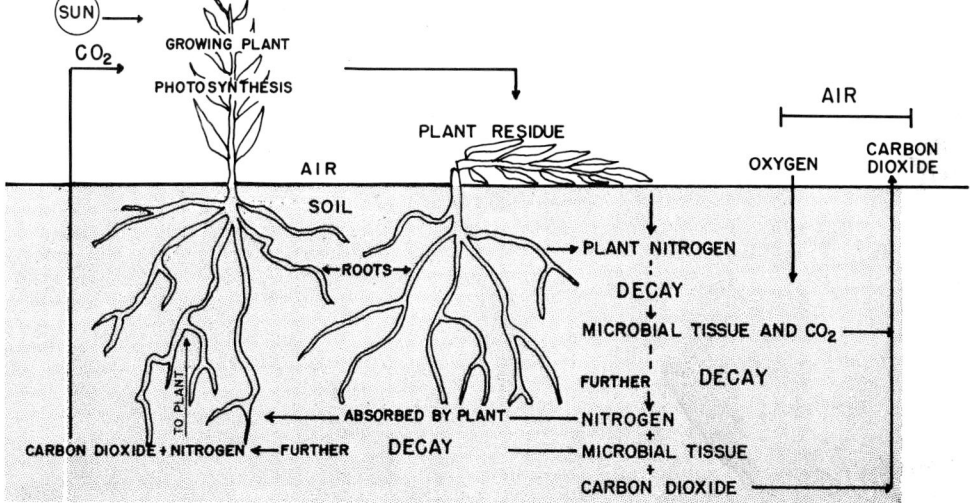

Fig. 2.5. Production decomposition and recycling of plant residues in soils.

physical benefits offset the organic matter consumed. Earthworms create no fertility since they do not fix atmospheric nitrogen or any other unavailable element and must live on the organic matter and plant residues found in the soil. Earthworms occur in desert soils whether irrigated or not. Seeding is not necessary. Wherever a food source and moisture are available, they appear in large numbers.

Benefit to physical properties. Individual soil particles are very small. If it were not for the fact that they tend to combine into aggregates (soil structures), plants would find it difficult to grow in soil, and water and air movement would all but cease except in coarse sands. Soil structure varies considerably, but not all structures provide optimum conditions for growing plants. Long experience and research have shown that those soils most abundantly supplied with organic residues (straw, manures, leaves, composts, etc.) possess the most favorable structures. A decline in soil organic matter is associated with a decline in desirable physical properties of a soil. Soils of poor physical condition or tilth are hard, cloddy, and break up with difficulty when worked with implements. Compaction in sub-surface horizons occurs when organic matter declines to a critical level; aeration is poor and water penetrates slowly. Plants under these conditions grow with difficulty. Desert soils of the Southwest lose organic matter rapidly upon cultivation and irrigation, even with the return of cultivated plant residues to the soil. Home gardens become shabby and disappointing to the homeowner when soils are allowed to deteriorate. Yield barriers, as well as poor quality in home gardens and urban landscapings, are due as much to the deteriorated physical condition of the soil as to any other single factor.

The importance of soil organic matter for optimum physical conditions in soils cannot be overstressed. Without organic matter and crop residues, the soil micro-organisms cannot produce the necessary bacterial products and organic residues to create good aggregation and physical condition. But the mere addition of organic residues to soil is not sufficient to produce good structure. It is the *continual* decomposition of organic matter, residues, etc., by the organisms with the formation of *new* products that is important. The right micro-organisms live in soils in sufficient abundance, ready to work on organic materials. To increase their activity, they require only an abundant energy source such as plant residues, other organic matter, and favorable moisture relationships. The principal way to insure good structure is to continually supply the soil with a source of organic food material for the soil organisms.

Decomposition

Organic materials entering the soil are subject to microbial decay. Even the most resistant of substances, such as asphalt, phenol, waxes, and resins slowly decompose as a result of microbial attack. A host of different substances decay when plant residues mix with the soil. Some of these are cellulose, hemicellulose, lignin, proteins, waxes, resins, vitamins, growth substances, and water-soluble compounds. The rate of decomposition is controlled primarily by the availability of the substance as food. For example, water-soluble constituents decay very rapidly, yet lignin is slowly decomposed. The more mature plant materials decompose less rapidly than younger materials, since lignin is more abundant in mature plants.

Maintenance

Maintenance of soil organic matter in desert soils is more difficult than in humid soils of temperate regions. The rate of decomposition in desert soil is rapid because of favorable temperature conditions most of the year (no refrigerative effect of frost and freezing). Maintaining even a low content of organic matter is essential to a favorable plant economy.

Maintenance may be illustrated by the example of a horse drinking from a trough of water. It is not the level of the water in the trough that concerns the horse as much as the rate of income of water in relation to his uptake. The trough may be four or 24 inches deep. If the water flows in at the rate he drinks, his thirst will be satisfied at either depth. However, if there is *no* income of water, the 24-inch depth would be the most desirable and would satisfy the needs of the horse longer than the shallow depth. Furthermore, the exploitation of the first 20 inches would be of little consequence, if replenishment at a rate equal to the outgo were begun and maintained at or above the minimum level for the horse's mouth to penetrate.

Desert soils of the Southwest, being lower in organic matter, need replenishment much more frequently with carbonaceous material than those soils higher in organic matter under a different climate less favorable to decomposition. Organic matter in soils may be maintained by the addition of any biological material if it is decomposable and is in sufficient quantity. The problem is to find organic materials that are inexpensive, but suitable. Agriculturalists use three main forms: (a) farm animal manures, (b) green manures, and (c) mature crop residues. Unlike the backyard gardener, in many farming circumstances it may not pay to use either (a) or (b) but, it would pay to get as much of (c) into the soil as possible. Fertilization, which increases the quantity of top and root residues available to turn back into the soil, definitely is a practice that all homeowners cannot afford to overlook. Another possibility is the use of municipal solid waste compost and feedlot animal wastes. The use of these sources appears to be developing as a result of over-burdened waste disposal facilities and pressure by the public for more effective waste control. In the final analysis, the renewal of organic matter in desert soils is largely dependent upon the utilization of all suitable organic residues possible. This is done for gardens by making a compost pile or spading it into the soil at the termination of growth. In both instances water must be added to keep the material damp to permit decay to take place. Mulching also can be practiced where it is convenient and not too unsightly.

Home gardeners and urban landscapers have a wide variety of organic materials from which to choose. These include the above, as well as bat guano, peat moss, compost, sawdust, shredded and lump tree bark, grass clippings, leaves, and straw, to name only a few.

SOIL WATER

The soil water or soil solution in deserts differs from that in humid regions primarily in its salt content, often expressed as parts per million of dissolved solids. The presence of salts makes it necessary to be alert concerning quantity and quality of water used for irrigation. The management of irrigation water in a desert climate is so critical that it has become a science. Slight changes in total salt content and/or in kind of salt (sodium, calcium, etc.) may make the difference between plant growth and no growth.

Quantity

Since rainfall is limited in arid regions, irrigation leaching must be accomplished to carry the salts below the root zone if harmful accumulation is to be prevented (Fig. 2.6). Dissolved salts in the soil solution move to the depth of water penetration. The salts in the soil solution come from both soil and water. In fact, irrigation water alone contains sufficient salt to put the land out of production if it is not leached occasionally. In home gardens and lawns, as in irrigated agriculture, leaching below the root zone should be practiced at least once a year. The amount of leaching water necessary to maintain productive soils depends on the quality or salt content of the irrigation and drainage water. The saltier the water, the more of it is required to keep a favorably low salt balance in soil.

Plants vary in adaptation to salt. For example, radish and green beans will tolerate 2,000 to 3,000 ppm of salts in the soil solution, whereas asparagus, beets, and spinach will tolerate three times as much. Soil water in humid climates contains only a few ppm of salts, rarely more than 200. Soil water in arid soils contains 10 to 100 times this amount.

Quality

Certain elements in irrigation water exercise a disproportionate influence on soil properties as compared with others. *Sodium* is one of these critical elements. If the ratio of sodium to calcium plus magnesium is high (i.e., above a ratio of 1:1), soil particles disperse,* water penetration and infiltration slow down, organic mat-

---

Disperse is used in soils science to mean that soil particles do not cling together but separate as individuals. Dust in the air is made up of separated (dispersed) particles. In a soil crumb structure that holds together, even when saturated, the particles are not dispersed but *flocculated*. Another example of dispersion is that of protein in milk, which is dispersed in sweet milk but coagulates (or flocculates) into a curd when sour and heated slightly (cottage cheese).

Fig. 2.6. Poor establishment of barley caused by inadequate leaching and movement of salts below the root zone.

ter solubilizes, and the black color of humus appears. Such conditions are referred to as *black alkali*. Black alkali soils are found under unirrigated-uncultivated as well as irrigated conditions. Soil water in sodic soils has a pH value above 8.4, is high in sodium, bicarbonate, carbonate, hydroxide, and phosphate. Only the most tolerant plants persist, if any at all. Sodium must be removed by deep leaching. Fortunately, only limited scattered areas of black alkali occur in the Southwestern deserts. The pH chart (soil reaction) shows the black alkali range as above 8.4 (Fig. 2.7). Most desert soils are not strongly alkaline, but range in pH from 6.8 to 8.4.

Some soils may be high in specific toxic ions such as boron. *Boron* is readily detected in the soil solution by chemical means and can be leached.

Some salt is necessary for good soil structure. For example, soils become dispersed when waters of very low salt content are used and the leaching of salts proceeds too far. Some salt is necessary to keep the clay flocculated. The large dense clods in Figure 2.8 formed when the soil dried after a very heavy rain following the seeding of cotton. Salts were washed out of the upper few inches, leaving the soil highly dispersed or deflocculated (puddled). The

Fig. 2.7. Soil reaction (pH chart) showing ranges of alkalinity common to desert soils.

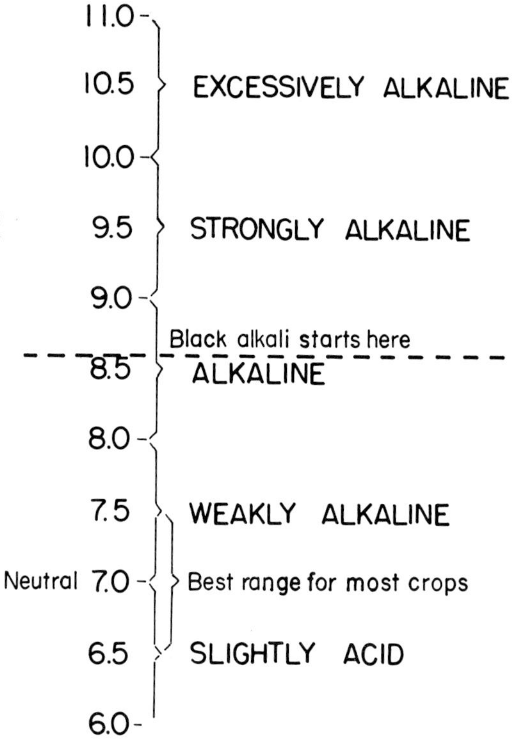

pH	
11.0	
10.5	EXCESSIVELY ALKALINE
10.0	
9.5	STRONGLY ALKALINE
9.0	
	Black alkali starts here
8.5	ALKALINE
8.0	
7.5	WEAKLY ALKALINE
Neutral 7.0	Best range for most crops
6.5	SLIGHTLY ACID
6.0	

Fig. 2.8. Soil highly dispersed with excessive shrinking and swelling as a result of too low salt content after intensive leaching by a heavy rain.

cloddy and caked surface prevented seedling emergence. In the South Gila Valley in the Yuma vicinity, Colorado River water was brought into the area and used in place of the more salty well water. The soil soon began to disperse and drainage was impeded. In both instances the application of salty well water corrected the problem. Mixing well water with Colorado River water is practiced on some porous sandy soils in the Yuma Valley to keep soil water at a favorable salt level, which serves to prevent harmful dispersion of the soil particles and extends the existing good water supply.

Water quality requirements for irrigation purposes are not the same as for domestic use. High quality domestic water may be wholly unsuited for irrigation and vice versa. Certain water softener devices, for example, use sodium chloride (table salt) to exchange with calcium to reduce water hardness. Sodium waters, on the other hand, deteriorate the physical condition of soils and eventually make them unsuited for growing plants. Compared with sodium, the hardness due to calcium does not detract from irrigation water quality. *Fluorides, chromates, lead,* and many other elements, whose presence makes waters undesirable for domestic sources, do not affect their value for irrigation.

The use of Colorado River water for irrigation has increased dramatically since the early years of the 20th century. The development of new irrigation districts since 1925 and expansion of old districts in the upper as well as lower basin states has resulted in some deterioration in quality of water all along the length of the river. As water passes from state to state there is a progressive increase in the salt content. Thus irrigators at Yuma and in the Imperial Valley can expect saltier water to enter their soils than enters those in Colorado, Wyoming, and Utah. Similarly, Mexican irrigators can expect to receive saltier Colorado River water than we in the United States. Had the earlier irrigators of the land known more about the soils they were cropping, and initiated better management practices, much of the salt problems would not exist today.

The concentration of salt in the water since the middle 1950s became an international problem between the United States and Mexico. However, in 1973 an agreement between the two nations was reached concerning the delivery of irrigable-quality water to the south. As part of this agreement, a desalinization plant is to be established near Yuma, designed to reduce the level of salt in the water being returned to the Colorado from irrigation projects in that area.

SOIL AIR

Soil is porous. The volume not occupied by solid soil particles is called *pore space*. If soil were Swiss cheese, the holes would be pore space. It is defined in terms of percentage of the total soil volume. Approximately 50 percent of the soil volume is pore space containing air and water. The main difference between soils of humid and arid climates is in the proportion of pore space holding air or water. Moisture occupies a greater percentage of the pore space for a longer period of time where rainfall is abundant than where it is scarce. Most soils are aerobic (they contain sufficient air to supply oxygen for favorable chemical and biological processes). Under certain high rainfall conditions, however, soils may be water-logged, i.e., the pore space is fully occupied by water for varying periods of time during which time anaerobic processes (not requiring oxygen) become dominant. Aerobic conditions favor root growth. Roots die in soils waterlogged for too long a time. Anaerobic conditions are less apt to form under arid than under humid conditions.

During the decomposition of plant residue and soil organic matter, oxygen is used by the micro-organisms and carbon dioxide is released. In uncompacted desert soils where vegetation is sparse and organic matter low, the demand for oxygen and release of carbon dioxide in the surface layer is lower than in soils of humid climates. Thus the proportion of oxygen to carbon dioxide generally may be higher in arid than in humid soils.

However, the small difference in composition of soil air between desert and humid soils is of little practical significance in plant growth as compared with other factors, such as temperature and moisture.

Differences between daytime and nighttime temperatures are greater in arid than in humid areas. Soil heat is lost to the atmosphere more rapidly at night in arid than in humid areas. Convection accounts for as much as half of the loss of heat from soil to air at night. Although surfaces of desert soils become hot, the soil is a good insulator. A few feet below the surface the temperature changes very little if at all between summer and winter. Plant roots therefore extend deep into cool soil to avoid the heat of the upper few inches. Animals find excellent air conditioning in deep burrows.

3. Why Soils Differ

Driving along the highway so many times we hear the comment, "That's a funny-colored soil."

Sure enough, a road cut has exposed a red, orange, or black soil, different from that in our own backyard. The soil may be deep, shallow, stony, or sandy. Layers of black organic matter may contrast against layers of white lime, or the soil may appear dull. Children who have a great curiosity for their surroundings ask the most embarrassing questions. One instance that comes to mind vividly happened on a soils tour into New Mexico. A soil specialist was explaining sand dune formation. A young girl blurted out, "Why is the sand here red and by the ocean where we went this summer it is white?"

Good question. How do we describe why soils differ so that the variations can be understood when we know they form under complex conditions?

Perhaps we can begin this way. Soils develop from the processes of weathering of many different kinds of rocks and minerals. During early geological eras the mineral particles were ground and worn from rocks by the erosive action of wind, water, and ice (Figs. 3.1 and 3.2). In many instances these particles have been moved great distances and mixed in many different proportions. Finally, new and secondary minerals, such as tiny clay particles, form from the weathering of primary minerals inherited from the original (igneous and metamorphic) parent rock. Thus, geological material becomes soil only after modification by what are called the *soil-forming factors*. Soils differ because these soil-forming factors differ, resulting in a great number of differing combinations of conditions.

[35]

ARID CLIMATE

HUMID CLIMATE

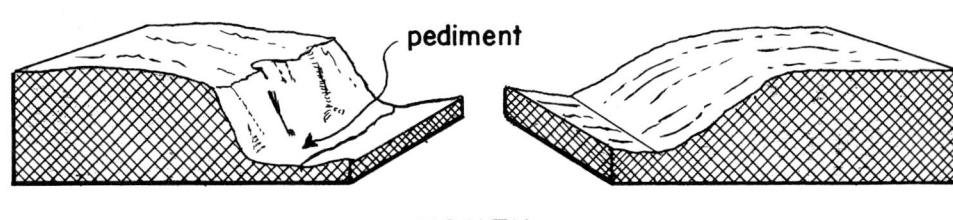

pediment

YOUTH

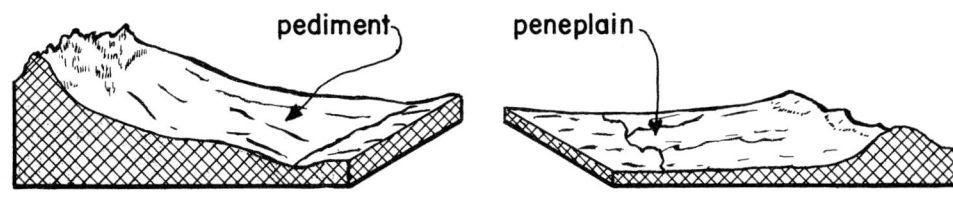

pediment

peneplain

MATURITY

Fig. 3.1. A diagrammatic comparison between young and mature topography formed under arid and humid climates.

Fig. 3.2. Badland erosion in the Painted Desert of Arizona.

SOIL-FORMING FACTORS

Factors which influence the processes that develop soils from rock and rock particles are *climate, vegetation, topography, parent material,* and *time.* The many varied interactions of these factors in affecting soil-forming processes (additions, removals, etc.) determine the characteristics of the *soil profile,* distinguishing it from its original mixture of heterogeneous inorganic minerals and/or parent rock. The term *profile* designates a vertical section through the soil, as shown in Figure 3.3. Some soils form *"in place"* over the ages from parent materials. Others develop from wind- or water-transported geological material. Wind-laid materials include *loess, sands* (*sand dunes*), and *volcanic ash.* Water-laid material occurs as narrow strips along streams and rivers and in arroyos and arroyo fans, and is called *alluvium.* Alluvial particle sizes range from large boulders to sands, silt, and clay.

One of the most fundamental changes brought about in geological material during the formation of soil is the accumulation of organic matter that is a product of the interaction between plant (and animal) residues and micro-organisms. Organic matter accumulates in soils after many years' growth of vegetation under favorable climatic conditions; arid soils accumulate it more slowly than other soils. This organic matter is one of the four main components of soil.

Briefly, these four main parts are:

1. *Mineral matter* derived from rocks but highly altered, composed of (a) particles which are separated for convenience into gravel, sand, silt, clay, and colloid fractions, and (b) accumulations of chemical substances such as carbonates (lime, dolomite), sulfates (gypsum), and oxides (quartz, hematite, and limonite).
2. *Organic matter* derived from plants, animals, and micro-organisms.
3. *Soil water,* which is a dilute solution of soluble salts and plant nutrients.
4. *Soil air,* which occupies about one-quarter of the soil space in well-drained soils.

Why do soils differ? They differ not only because geological materials from which they form differ, but because the hundreds of species of minerals become altered and redistributed during weathering, and new minerals form. Moreover, the plant cover is not the

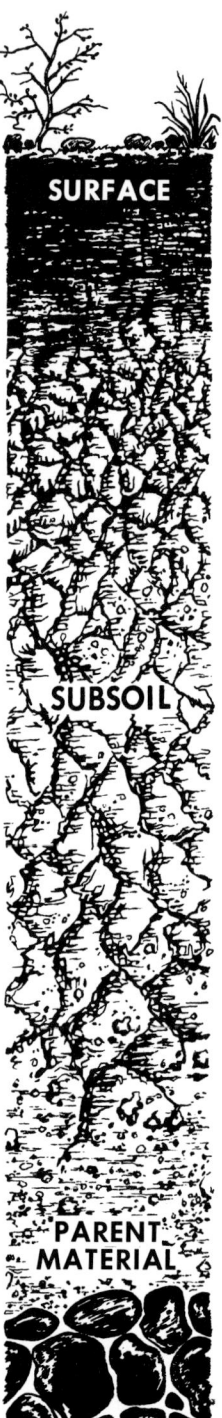

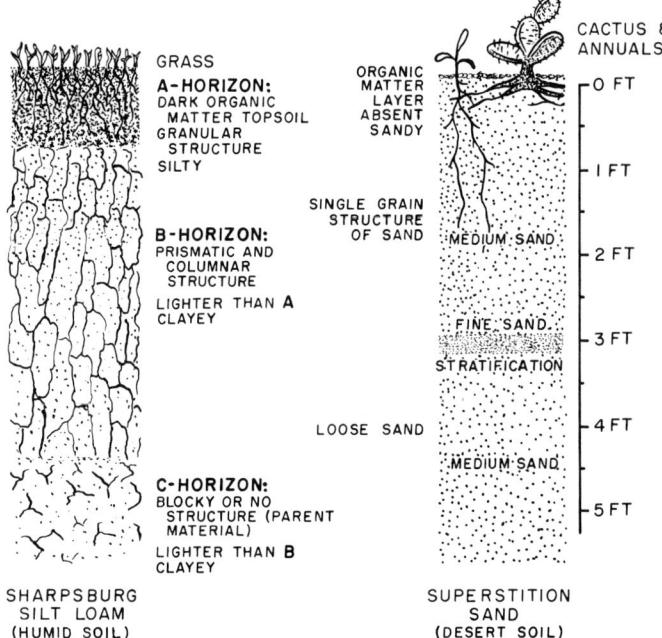

Fig. 3.3. The soil profile is a vertical cross-section of a soil. There are three general horizons: surface, subsoil, and parent material. Bedrock or geological material of varying composition lies below the horizon.

GRASS

A-HORIZON:
DARK ORGANIC
  MATTER TOPSOIL
GRANULAR
  STRUCTURE
SILTY

B-HORIZON:
PRISMATIC AND
  COLUMNAR
  STRUCTURE
LIGHTER THAN **A**
CLAYEY

C-HORIZON:
BLOCKY OR NO
  STRUCTURE (PARENT
  MATERIAL)
LIGHTER THAN **B**
CLAYEY

SHARPSBURG
SILT LOAM
(HUMID SOIL)

ORGANIC
MATTER
LAYER
ABSENT
SANDY

SINGLE GRAIN
STRUCTURE
OF SAND

LOOSE SAND

CACTUS &
ANNUALS

MEDIUM SAND

FINE SAND
STRATIFICATION

MEDIUM SAND

SUPERSTITION
SAND
(DESERT SOIL)

0 FT
1 FT
2 FT
3 FT
4 FT
5 FT

Fig. 3.4. Diagram of two profiles showing Sharpsburg silt loam, a typical prairie soil, and Superstition sand, a desert soil also developed on wind-deposited material.

same on all soils. Land slopes and microclimate differ. All soils, therefore, do not have the same level of natural fertility and productivity.

Soils may be shallow, or deep. These are relative terms, but most people think of a shallow soil as one where rock is struck within spade depth. Alluvial soils on valley floors, benches, terraces, or fans are generally deeper than soils on slopes where they are subjected to erosion and must develop in place from consolidated rock or rock material.

A soil has distinct characteristics which appear in layers called *horizons* approximately parallel with the surface. These horizons may be seen in the field by exposing a cut or vertical section of the soil as diagrammed in Figure 3.3. Compare the typical cool-humid soil with the typical desert soil of the Yuma mesa (Fig. 3.4). The *soil profile* is an exposed section of soil including the surface organic layer, the layers directly below, and the parent material or other layers beneath that influence the formation and behavior of the soil. Not shown in the arid soil diagram are other distinctive layers which often occur, such as *lime, gypsum, iron cementations,* and *clay.* When compact, these layers are referred to as *pans, hardpans,* or *duripans.* They often seriously hamper root penetration of plants. Usually these pans are located within or at the bottom of the B horizon, or accumulation layer.

CLIMATE

Climate has great influence on the characteristics of soils. In a desert such as the Sonoran, temperature and moisture (rainfall or irrigation) dominate the processes which differentiate soils. For example, if the soils of earth were as completely devoid of moisture as is the moon's surface, only physical weathering would be possible, assuming the present ranges of surface temperature, and soils would differ very little, if indeed they would develop at all.

Temperatures

Differences between day and night temperatures in the Yuma subdesert range from 30 to 40 F, far greater than those of humid climates. During the day the desert soil absorbs more solar heat than that of humid lands, and at night more heat escapes to the atmosphere. Nights turn cool, whereas in wetter climates they remain more moderate. Farm crops can be grown throughout the year if provided irrigation. Temperature, however, influences the rate of accumulation of soil organic matter. Even with low rainfall, organic residues decompose rapidly after scant rains. Unlike colder climates, only short periods of refrigeration occur during the winter months, even in the higher Sonoran subdeserts. Decomposition takes place virtually throughout the year. The dark organic layer, found so commonly in forested and grassland soils, is not evident in soils formed under the present desert climate.

Clay forms more slowly under arid than humid conditions. Clays in the Sonoran desert were formed during periods of humid climatic cycles, or were transported in by wind and/or water action.

Moisture

The annual rainfall at Yuma and in the Imperial Valley averages slightly above three inches. In other subdeserts in the Southwest the annual rainfall ranges up to 11 inches. Most of it falls rapidly. Soil washing, erosion, and runoff are intense. The high runoff rate further reduces the rain's effectiveness for plant growth except along stream channels, arroyos, and valleys where water accumulates. Shrubs and trees grow more densely along these wetter drainageways, and soils show the effect of more organic matter.

VEGETATION

The scarcity of vegetation limits the amount of residue available for soil organic matter production in arid climates. Since nitrogen is carried in soil organic matter, it is low in desert soils. Because of the low organic matter content, the surface color of desert soil is altered little by humus. The topsoil and subsoil may be of similar color. Unlike Midwest prairie soils, the richness or fertility of a desert soil cannot be assessed on the basis of a dark surface. In fact, red colors correlate more with fertility or productivity in Southwestern desert soils than dark colors.

Recognizing that vegetation in a desert is controlled by climate eliminates climate as an independent variable in soil formation. This does not mean that vegetation does not affect the soil. In fact, certain species of desert plants accumulate soluble ions such as sodium salts. The soil located under and in close proximity to these plants may take on a wholly different physical character. It may be highly alkaline and dispersed. When the land is cleared for irrigated agriculture or homesites, these small areas support plant growth poorly, if at all, and absorb water slowly. They must be given special reclamation treatment, which often is time consuming and expensive.

Vegetation influences desert soils in ways other than by accumulating salts. Early in the history of nursery growing in the Southwest, soil material for containers was collected from under scattered desert shrubs and trees. Nurserymen especially sought the rich organic soil accumulated under the "limb-line" of mesquites. The nurseryman took advantage of its greater fertility. The presence of organic matter and a desirable granular structure favored rapid growth and good root development of container plants. This practice is diminished now because of the exhaustion of nearby supplies.

Driving through deserts of the Southwest, one can see great variation in the extent of shallow sand drifts. Shrubs, particularly creosote bush and burro bush in low and open stands, collect dune sand (Fig. 3.5). As sand continues to pile up around these plants,

Fig. 3.5. Sand dunes partially covering creosote bush on the mesa above the Rio Grande west of Las Cruces, New Mexico.

they continue to grow, depending on the particular wind pattern. It is not uncommon to see some plants nearly covered with several feet of sand. When the dunes are leveled, plants removed, and the soil irrigated and planted, variations in plant growth in the dune areas are readily distinguishable from those in the adjacent soil. The soils beneath these dunes contain different levels of fertility. Some are poor in nitrogen and others in phosphorus and trace elements. Differences in natural vegetation along narrow drainageways in alluvial valley soils, as compared with vegetation on higher ground, also are commonly observed. Again, when the land is leveled and planted, textural stratifications and physical differences often remain and growth is quite irregular.

TOPOGRAPHY

The Southwest desert land lies primarily in alluvial valleys and on terraces and higher mesas. Topography has a major role in desert soil differentiation. Soils of depressions, which receive runoff and drainage from higher levels and accumulate salts, are different from adjacent soils. A transect of soils from the San Joaquin Valley illustrates this principle so well it is used here, even though this desert falls outside the Sonoran desert boundaries. Soil series (Pond and Fresno) of the valley floor position are much higher in salinity (soluble salts) than those on slopes (Cajon and Traver, Table 3.1). Soils of the Cajon and Traver series have some areas relatively free of salt, whereas the Pond and Fresno series are highly saline with none of the land free of salinity.

TABLE 3.1

Influence of topography on the salt content of the Fresno Family of soils.*

Soil Series	Topography	No alkali <0.20%	Slight alkalinity 0.2-0.4%	Moderate alkalinity 0.4-1.0%	High alkalinity >1.0%
Cajon	Sloping	41	22	16	22
Traver	Sloping	10	34	37	19
Pond	Level (valley)	0	3	29	67
Fresno	Level (valley)	0	0	7	93

*Distribution of salinity (% of total area)

*Data compiled by D. F. Foot from unpublished maps of Soil Survey of the U.S. Department of Agriculture and the University of California, as published by Hans Jenny in *Factors of Soil Formation* (McGraw-Hill, New York, 1941).

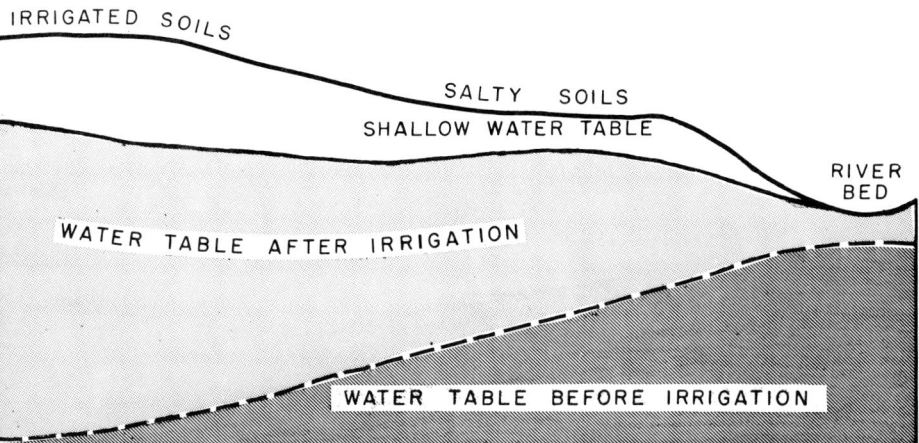

Fig. 3.6. Illustration of salt accumulation in soils as a result of man's irrigation practice in a situation where topography and water table height can interact unfavorably.

Where textural stratification of alluvial material occurs and infiltration is low, salts may be brought to the surface by capillary rise and accumulate upon evaporation of moisture.

Irrigation practices may also contribute to salinity as a result of topographical differences of soils. In the early 1930s, the area surrounding Buckeye, Arizona, was almost "salted-out" of plant production by irrigation water applied on higher land, which flowed laterally to the lower level and accumulated as a high water table (Fig. 3.6). Salts leached from soils of the higher area moved up by capillary action from the shallow water table to the surface. Crop production deteriorated and much of the land required reclamation to put it back into economic production. Reclamation practices involved washing out the salt with good water after applying gypsum and providing suitable drainage for the leachate. Yuma Valley in the South Gila Project experienced a similar problem when Colorado River water was brought to the mesa for irrigation purposes.

Imperial Valley growers are able to discharge the salt leachate from their soils into the Salton Sea. Until they began irrigation with Colorado River water, the depression now filled by the sea was dry. Adequate drainage of soils is critical in maintaining productive soils in arid regions. Small topographical changes have greater influence on the plant growth characteristics of arid soils than upon those of any other climatic region.

Erosion of soils on even the gentlest slopes can be serious under desert conditions, partly because soils neutral-to-alkaline disperse readily, and partly because the intensity of rainfall usually is greater than in most other regions of the United States. Furthermore, vegetation is sparse, leaving much of the soil bare and exposed to direct erosive action of rain. Topography and climate interact as major factors influencing soil profile characterization.

PARENT MATERIAL

Parent material is recognized as the initial state of a soil system. It may be either rock or weathered rock, in place or transported, according to where one wishes to begin in the development series:

Rock (consolidated) → weathered rock (unconsolidated) → immature soil → mature soil

Much of the desert land of the Southwest has been deposited by water or wind. The wind-laid materials originally were blown out of adjacent alluvial deposits located along river or stream beds. The

TABLE 3.2

Percentage chemical composition of some rocks (from United States Geological Survey)

Oxides	Igneous rocks	Shales	Sandstones	Limestones
SiO_2	59.14	58.10	78.33	5.19
Al_2O_3	15.34	15.40	4.77	0.81
Fe_2O_3	3.08	4.02	1.07	0.54
FeO	3.80	2.45	0.30	. . . .
MgO	3.49	2.44	1.16	7.89
CaO	5.08	3.11	5.50	42.57
Na_2O	3.84	1.30	0.45	0.05
K_2O	3.13	3.24	1.31	0.33
H_2O	1.15	5.00	1.63	0.77
TiO_2	1.05	0.65	0.25	0.06
CO_2	0.10	2.63	5.03	41.54
SO_3	0.13*	0.64	0.07	0.28*
P_2O_5	0.30	0.17	0.08	0.04
MnO	0.12	. . . .	. . . .	0.05
Others	0.28	0.05	0.05	. . . .

*Includes elemental sulfur calculated as the oxide.

starting point of soil, as Jenny* puts it, "state of soil at soil formation time zero," is weathered rock. Thus the starting materials of
valley and mesa soils have a common origin. The wind-deposited
material, though, has undergone resegregation in the blowing process. Compared with river alluvium, aeolian material is relatively free
from vertical stratification of sands, silts, and clays. In this manner,
parent material differences between the two locations resulted from
an interaction with climate (i.e., wind).

An outline of the broad classes of parent material and the
chemical composition of some rocks (Table 3.2) is provided as a
guide for a better understanding of the differences between soils and
the materials from which they form:

1. Residual rock
    a. Igneous — granite, basalt, and andesite
    b. Sedimentary — limestone, sandstone, and shale
    c. Metamorphic — marble, quartzite, and gneiss
2. Transported material
    a. Water
        Alluvial — running water
        Lacustrine — lakes
        Marine — ocean
    b. Wind
        Sand dunes and very fine dust
        Loess — primarily silt-size particles
        Volcanic ash
    c. Ice (glaciers)
        Moraine
        Till plain
        Outwash plain
    d. Gravity
        Colluvial
3. Cumulose material
        Peat and muck

Physical weathering (as illustrated in Fig. 3.7) predominates
in arid climates. Soils of appreciable depth in the arid Southwest form
very slowly if at all from native rock under the present limited rainfall conditions.

*Hans Jenny in *Factors of Soil Formation*. McGraw-Hill, New York, 1941, p. 99.

Since soil formation is so slow under these conditions, the characteristics of the soils are more directly related to the characteristics of the parent rock material from which they are derived than soils in humid climates where weathering and soil-forming processes occur at comparatively accelerated rates.

Sand dune topography provides an interesting study of both salinity and plant distribution when compared with irrigated conditions. Under natural conditions salinity and vegetation in large sand dunes may be expected to distribute according to the rainfall pattern of frequency and intensity. The least salt occurs at the peak of the dune and the most at the bottom trough between dunes. Leaching has obviously moved the salts down into the trough. The least salt-sensitive plant species grow at the top, whereas no growth takes place in the trough.

This natural example should not be confused with the furrow irrigation condition of the microtopography of shallow beds and side-ridge planting where the reverse condition exists (Fig. 3.8). Salts accumulate by capillary rise at the highest rather than lowest point. The soil in the trough leaches until the salinity is reduced considerably below that of the ridge.

Fig. 3.7. Exfoliation of rock in the process of physical weathering.

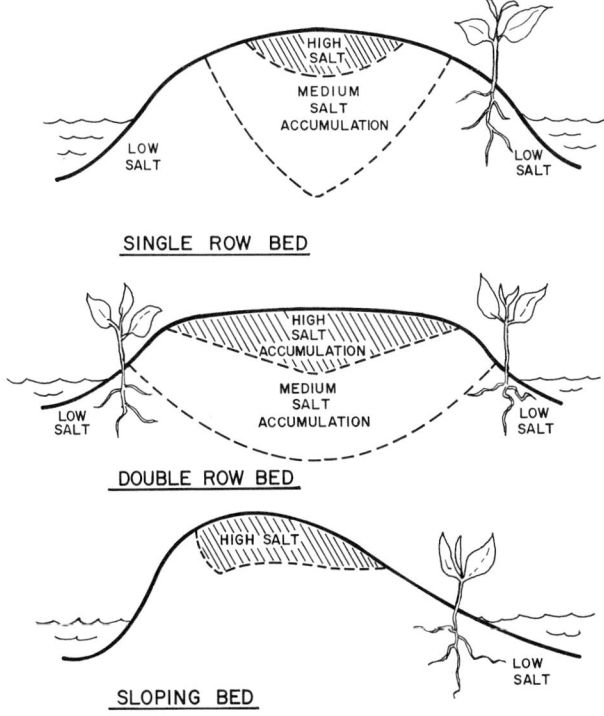

SINGLE ROW BED

DOUBLE ROW BED

SLOPING BED

Fig. 3.8. Influence of bedshape on the accumulation of salts. The placing of the plant is critical to its survival.

TIME

Time is related to soil formation in arid soils as an interacting function of climate. The importance of moisture in soil formation is so great that soils of arid lands in the Southwest develop very slowly. The influence of water and vegetation on accelerating soil development may be clearly demonstrated by comparing residential or golf turf lying adjacent to virgin desert. In a very few years the soil under the turf sod develops a distinctly dark organic layer, and carbonates move down to concentrate at the greatest depth of leaching. In finer-textured soils, i.e., loams and clays, a compact layer often develops below the organic accumulation layer. Under agricultural field conditions, soils such as Mohave sandy loam accumulate lime (carbonates) in the upper foot, although under virgin conditions no carbonates were present. The carbonates come from the irrigation water that is usually well-supplied with the ingredients of lime. Organic matter accumulation is not pronounced in garden or field-cropping conditions. Compaction of the subsoil is a common occurrence in almost all soils where traffic is necessary and excessive.

4. What the Surface Looks Like

In traveling from one area to another, changes in the appearance of the soil surface become a prominent feature of interest. Once in Argentina I heard a fellow traveler say, "That soil looks black like the prairies of Iowa." A newcomer to Arizona was heard to ask, "How can anything grow in this sunbaked adobe and dune-sand desert?" Thus soil surface peculiarities enter into observations of our curiosity. This chapter is concerned with the appearance of desert soil surfaces which differ markedly from those of other climates and deeply influence living in the desert. It contrasts to the following chapter which familiarizes us with what to expect below the surface.

The most striking feature of desert land surfaces is its comparative lack of plants. Soil color, texture (sand, silt, clay, gravel), surface structure (granular, paved, blocky, loose, etc.), and rocky or stony surfaces are visible on desert terrain because of sparse vegetation. Topographical changes (dips, swells, ravines, knolls, sinks, etc.) also are more apparent in the desert where vegetation grows sparingly. The exposed ground surface of the desert Southwest ranges between 20 to 60 percent compared to zero under humid conditions. Thus, soil surfaces become more a part of the inhabitant's life in arid than in other climates. Quality clays and stone are easily discovered and readily accessible for use in desert lands and provide excellent construction materials, as evidenced by utensils, artifacts, and buildings. Adobe bricks, made by drying and baking clay in the sun, provide building materials at a minimum cost (Fig. 4.1). Anyone who can find a clay pocket can build an adobe house.

COLOR

Dark surfaces, indigenous to areas where rainfall is sufficient to grow trees, shrubs, and grass in abundance, are lacking in desert soils. Organic matter and humus do not accumulate in quantities great enough to give desert soil a dark color. The surface colors of

Fig. 4.1. Adobe brick baking in the sun makes excellent building material at a minimum cost.

deserts, therefore, relate more to various states of oxidation of minerals than to any other factor. Iron oxides vary widely in color from dull gray-yellow to rougelike red. The iron oxide compounds listed below give distinctive color characteristics, depending on the state of chemical combination and hydration.

Hematite	Fe_2O_3	Red; hydrates with an increase in volume to form yellow limonite.
Magnetite	Fe_3O_4	Dark brown; magnetic; resistant to change and not widely distributed.
Limonite	$2Fe_2O_3 \cdot 3H_2O$	Yellow; soft; resistant to change except in extent of hydration, with wide distribution.
Others	$2Fe_2O_3 \cdot H_2O$ $Fe_2O_3 \cdot H_2O$ $Fe_2O_3 \cdot 2H_2O$	Yellow-red to yellow with increase in hydration (H_2O). These provide much of the reds and yellows in soil color.

The Mohave soil series* of Arizona and California, for example, possess a distinctive red to reddish brown color, due primarily to iron (as hematite and magnetite) in a peculiar oxidized state. As much as 12 percent of the soil by weight can be calculated as iron oxide.

*Soil series names originate from towns, cities, or smaller communities where the soil was originally discovered and described.

The pale gray of the Laveen series, often found near the Mohave, gets its ghostly color from lime, which imparts a gray and white coloration to soil. Gypsum and quartz are white or clear. Gypsum sand as found at White Sands National Monument in New Mexico is an example where at first sight, it is difficult to believe the sand is gypsum and not quartz. Manganese oxides range from black to brown, and mica is silver, black, or brown. When a variety of such minerals, and others too numerous to include here, occupy the soil, a wide spectrum of colors is possible. In fact, soil scientists may refer to the "red desert" of the Southwest and the "gray desert" of the colder central and northwestern United States, when they are speaking of these soils in general.

Landscape planners of turfs and gardens prefer "red mesa" soil to other kinds because they know the red color is associated with good drainage, low lime (or caliche), and low salts. Color is an important factor in choosing a productive soil, whether it be the black humus of humid climates or the red of the arid Southwest. In general, the surface color of desert soils tends toward paler and softer shades than soils of humid regions.

TEXTURE

The mineral part of soil is conveniently separated into three broad classes of particle sizes — *sand, silt,* and *clay,* referred to as textural classes. These classes relate to the size of the individual minerals which make up the soil (see Table 4.1). Sands represent the largest particles, i.e., fine sand 1/4 to 1/10 mm and coarse sand 1 to 2 mm in size. Clay particles are the smallest, i.e., less than 0.002 mm, and silt sizes lie between sand and clay. When one textural class predominates, the soil is called *sandy, silty,* or *clayey.* Soil textural classes are based on the surface texture and not subsurface texture. In some instances surface and subsurface textures are alike. More often they are not. When gravel or stones occur in abundance, the term *gravelly* or *stony* is added as a modifier to sand, silt, or clay, i.e., gravelly clay or gravelly clay loam. Soils having about equal proportions of silt and clay and 50 percent or more sand are called *loams.* Sandy soils dry more quickly, drain faster, and need more frequent watering to keep plants alive than fine-textured clay soils. They not only hold less water than silt or clay, but also have less

TABLE 4.1

The soil particle size

Soil particle class	Diameter range	
	Millimeters*	Inches
Very coarse sand	2.0 –1.0	0.07874 –0.03937
Coarse sand	1.0 –0.5	0.03937 –0.01969
Medium sand	0.5 –0.25	0.01969 –0.00985
Fine sand	0.25–0.10	0.00985 –0.00394
Very fine sand	0.10–0.05	0.003937–0.00197
Silt	0.05–0.002	0.001969–0.00078
Clay	$<$0.002	$<$0.00079

*1 millimeter = 0.03937 inch.

See Appendix B for conversion table.

surface area and exposed chemical bonds to hold plant nutrients. Because water moves through sands rapidly, salts accumulate less than in clays. Soils of the desert Southwest contain a greater proportion of sand than those in humid climates. The sand also has sharper edges and is grittier than that east of the Mississippi River.

The finest particles are clay. The rate of water infiltration and aeration is slow in clay soils. On the other hand, clay has a greater water-holding capacity and higher plant nutrient storage level than sands. Silt is between sand and clay in these properties. Soil organic matter is so finely textured that it is colloidal in size like clay and has the highest nutrient storage level.

Most soils possess a combination of textures (Table 4.2) and are grouped into classes depending on the dominant distribution of particle sizes. Appendix, Figure A.5 gives a chart of texture sizes and the corresponding classes. The most desirable "textural classes"

TABLE 4.2

Percentages of sand, silt, and clay in some common soil classes

Soil class	Sand	Silt	Clay
Clay	$<$45	$<$40	$>$40
Clay loam	20–45	15–50	25–40
Silt loam	$<$50	50–90	$<$30
Loam	25–50	30–50	5–25
Sandy loam	45–85	$<$50	$<$20

for growing plants in arid climates are the loams: sandy loam, loam, silt loam, silty clay loam, and sandy clay loam. A texture of 10 to 20 percent clay and no more than 30 percent coarse sand is ideal for most plants. Garden soil should be free of stones, be neither sand nor clay alone, and have no compact or indurated layer of lime or gypsum within the most active root zone.

Texture, unlike structure, cannot be changed appreciably by cultivation or working. Where deep plowing (30 to 42 inches) mixes layers of different particle sizes and brings sand or clay up to the surface, the textures of the soil can be redistributed and made more uniform. This practice results in a more desirable medium (texturewise) for plant growth, aeration, and water infiltration.

Clay

Clay deposits ranging in thickness up to several yards may be found as surfaces in plains and depressions of the deserts. Clay plains are relics of a more humid climate and are often truncated soils or old lake or sea bottoms. Clay depressions, such as playas and old clay-deposited lagoons (Fig. 4.2), are almost impossible to put into

Fig. 4.2. A clay lagoon in a river bottom showing sparse growth of grass despite attempts to till and seed it.

Fig. 4.3. Moving sand dunes near Palm Springs, California.

agricultural production. They make very poor building sites because the swelling and shrinking unstabilizes foundations. The density and dispersion of clay is not amenable to simple corrective measures. Water penetrates slowly, aeration is poor, and roots do not penetrate to any significant depth except for scattered, hardy native plants of low economic value.

Clay streaks, islands, lenses, and layers may appear in river channel soils, in alluvium, and in stratified valley materials as barriers to the downward movement of water. These need special drainage treatment to be productive. Deep plowing or ripping to break up and redistribute these layers aids materially in more favorable air and water movement.

Sand

Deserts call to memory *sand, sun,* and *wind.* Although all deserts are not wastelands of shifting sand dunes, as depicted in movies with a French Foreign Legion setting, sand is a common desert constituent. Wind blows sand about in the desert in a restless way (Fig. 4.3). Blowing sand causes serious shearing off of vegetation at ground level (particularly if plants are in the young seedling stage)

Fig. 4.4. Rough rock land, typical of much of the upland in the desert Southwest.

and damages leaf crops, blooms, and young, tender fruits. Creeping sand dunes may bury valuable cropland, and can keep home yards in a ragged condition, encroach on park and golf turf, and even obscure streets and homesites.

Sandy soils must be watered more frequently than finer textured soils because they hold less water. They also require fertilization more often, particularly with nitrogen, since rapid movement of water washes soluble nutrient elements out of the root zone. Some sands are so coarse and droughty that it is extremely difficult to establish home gardens and flowers during summer months. This can be relieved somewhat by incorporation of peat moss, compost, grass clippings, and other organic residues, or by bringing in clay soil for mixing.

Gravel, cobbles, and stones

Rough, stony surfaces appear on a variety of desert topographies, most prominently on stream outwashes, sediment plains, alluvial flood plains and fans, and arroyos, and on consolidated uplands as foothills, denuded hills, lower mountain footslopes, and rough, mountainous land (Fig. 4.4). Rockland is used, agriculturally, for pasture and grazing. Homesites on accessible rockland can be beautifully landscaped and planted. Small rocks may be used as a mulch by placing them on strategic slopes to reduce erosion. Where the soil is shallow, tree-planting holes must be excavated and filled with good red desert soil.

STRUCTURE

Structure refers to the *arrangement* of soil particles of different sizes bound together in a unit. It is the architecture of the soil. Soil structures range from single particles, such as a grain of coarse sand, to massive clods of clay (see Fig. 4.5). The "crumb" structure that occurs in loamy soil under grass sod is ideal for growing plants. Desert soil surfaces have soft structures. The thin, platy surface is underlain with a vesicular (very fine air channels) structure formed by escaping air bubbles following wetting by infrequent rain. Alluvial valley soil structure often is transitory since it is only slightly stable upon wetting. It is practically impossible to walk on most desert soils after they have been irrigated without sinking to the depth of water penetration.

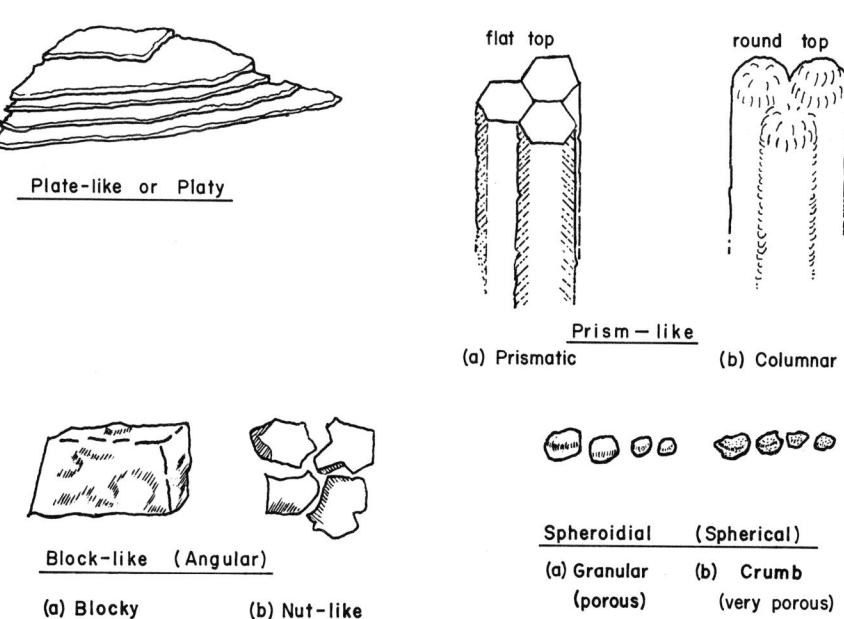

Fig. 4.5. Various types of soil structure are found in soils of the desert Southwest.

Fig. 4.6. Barren soil showing a massive blocky structure as a result of working it while too wet during a rainstorm followed by excessive irrigation.

Sands form aggregates poorly, if at all.

Puddled is a term referring to soils that have been worked when too wet. The soil structure has broken down and the particles have dispersed into a close compact single-grained arrangement. Water movement is seriously impeded or even prevented in puddled soils. Puddling is most serious with clayey soils. Upon drying, puddled soils form hard lumps and clods, reduce plant stand by inhibiting seed emergence, and limit root penetration and aeration (Fig. 4.6). The gardener often calls them "baked" or "caked." And indeed they are, for the ancient process of making adobe brick for building purposes (still in practice) is to pour water into a clay puddle or basin, knead the mixture with the feet while wet, and pour the puddled clay soil into frames for drying into bricks in the sun. Tilling clay soils does not result in puddling and clod formation if done when the soils are not too wet.

Soil structure can be improved for plant growth by adding animal residues and plant material such as grass clippings, straw, compost, sawdust, and green manures. Good structure is important. It controls water relationships, aeration, microbial activity, and root penetration in soils. Structure also may be favorably modified by the addition of certain amendments like gypsum and sulfur, by management of the wetting and drying cycle, and by rotation with certain soil-building crops. Organic polyelectrolytes have been used to stabilize good structure but they are quite expensive, even to the home owner. They all but disappeared from sales shelves after the 1950s.

PAVEMENTS

Wind and rain action may carry away fine particles of broad desert areas, leaving gravels, cobbles, and stones which are exposed finally in sufficient density to appear as pavement. These stony residues may become cemented together with lime, gypsum, and silicates in such a way as to further give the impression of a man-paved surface. "Desert varnish" also may develop on these stones from coatings of dehydrated iron hydroxides and manganese oxides polished by the action of wind-blown sand. Desert pavements may be seen from Ajo to Gila Bend, Arizona, and northwest in patches as far as Needles, California (Fig. 4.7).

Fig. 4.7. Desert pavement caused by wind blowing away the fine particles, leaving the pebbles to form a surface characteristic of road paving.

Fig. 4.8. Stones reflecting the high polish known as desert varnish.

EROSION

Ironically, the constituent most lacking in the desert — water —
plays an excessive role in shaping the physical characteristics of its
surface. The sparse plant cover exposes the land to spectacular erosion, as seen by scarred slopes, buttes, escarpments, deep winding
and twisting canyons, and stone- and boulder-strewn arroyos. The
erosive influence of brief but torrential local rains leaves land surfaces of strewn pebbles and angular stones, gullies and deep-cut rills,
and salt flats of thick clay that bakes and cracks in the heat. Water
and wind work the raw plains of the desert more violently than those
in other climates. Sand blasts whip against gullied hillocks, spires,
and jagged peaks, and relentlessly move dunes through mud flats
and basins. Dust storms follow dry stream beds and spill over towns
and cities which only recently have expanded along the banks and
terraces into the rocky foothills. Uninhibited wind collects sand and
sends it moving again and again over the land to form loose crescent
dunes, windrows along fences, and mounds around scattered stands
of yucca and creosote bush.

Yet the eroded materials collect on soil surfaces and develop
into fertile soils in the valleys, terraces, and mesas along streams and

rivers. These areas are represented by the lush citrus groves and vegetable fields which may extend as far as the eye can see. The rich soils are planted two or three times a year, yielding high value cash crops. Exotic landscapes of prosperous towns and cities continue to develop as immigration explodes into the desert. The surface of these level and loamy lands has required stabilization by soil conservation practices to counteract the tendency for serious erosion (Fig. 4.8).

TEMPERATURE

Although we actually cannot *see* the temperature of the soil surface, we frequently hear someone saying, "it looks hot." It *is* hot. In those months between rains, and when the intensity of the sun increases to its peak in June, soil surface temperatures may well exceed 125 F all day. Stones become so hot they cannot be held in the hand or even picked up without the hand being burnt. Seldom do we see children walking or playing barefoot on exposed soil. Even decks around swimming pools are covered with insulating "cool deck." Yet a few feet below the surface temperatures remain pleasantly cool throughout the year. The soil is an ideal insulator. The great loss of heat by convection to the atmosphere helps to make soil surfaces cool at night. The great temperature highs during the day and lows at night are one of the distinguishing marks of the desert.

5. What the Subsurface Looks Like

Smell the fresh earth crumbling under the spade's turn. It is the womb for seeds' new birth.

How many of us know what the soil in our yard looks like below the surface? Yet the subsurface is the plant's birthplace, store-house, and dining room. Between one third and one half of the plant lives there. It contains nutrients, water, and air (which roots need to breathe), and acts as an anchor to keep the top upright.

Since the subsurface is the feeding area of plants, its size and quality for root extension determine the health and vigor of the plant. It is comparable to pasture for the cow or range for the steer. If, for example, the root area is small because of rocks, shallowness, caliche, and hardpans, the *quantity* of plant nutrients may be limited, just as an animal confined in a small pasture finds limited feed. If the root-feeding zone is too sandy, low in nutrients and moisture, the *quality* of the root-feeding area is limiting, even though no physical barrier is present. Again, if the soil is salty and/or high in sodium or toxic boron salts, the root-feeding zone is limited quality-wise so that growth is stunted or the plant is unable to survive.

As soils develop from parent rock, such as granite, many different elements are released to form soluble salts. These salts combine in various ways with other mineral constituents and molecules of air and water. They accumulate in the soil of arid lands where the rainfall is not sufficient to wash them below the roots. These and other prominent desert *subsurface* characteristics which affect plant growth are discussed here, since they must be dealt with in home plantings and landscaping as well as in commercial food and fiber production.

PARENT MATERIAL

Igneous, sedimentary, and metamorphic rock are classes of source materials from which soils form. Each influences soils in a different way. Within these groups, a wide variation in composition is also reflected in the soil. For example, *residual* soils (soils formed in place) originating from igneous rocks, granites, rhyolites, basalts, etc., differ, just as soils originating from sediments, sandstone, and limestone do. In desert areas many soils develop from unconsolidated sediments that are grouped according to transport agencies — wind, water, or ice, technically designated as aeolian, alluvial, and glacial respectively. The subsoil which plant roots encounter thus can be quite variable. It may be deep or shallow, depending on how readily the parent rock or rock material breaks down by weathering. A soil derived from sandstone obviously is sandy just as a soil from limestone or granite is clayey. The texture of the subsoil thus reflects these differences. On the other hand, some sandstones become so highly weathered the surface is sandy and the subsoil is clayey since the fine clay moves down in the profile more easily than the coarse sand. However, different parent material is only one reason that subsoils vary as potential rooting media.

SOIL PROFILE

The soil profile is exposed by digging a pit down through the soil into what is called parent material. Profiles inform us about some of the individual soil characteristics influencing root penetration and plant growth, such as subsurface obstructions, salt concentrations, rocks, hardpans, shallowness, texture, and structure. A *soil* is difficult to define. Though the following suggested description admittedly is not adequate, it may be helpful in gaining a concept of a soil. The United States Soil Conservation Service states,* "We may say that it [soil] is the collection of natural bodies on the earth's surface, supporting plants, with a lower limit at the deeper of either the unconsolidated mineral or organic material lying within the zone of rooting of the native perennial plants; or where horizons impervious to roots have developed, the upper few feet of the earth's crust having properties differing from the underlying rock material as a result of interactions between climate, living organisms, parent material and relief." Soil is not just the upper humus-rich layer nor just any unconsolidated (loose) material regardless of depth or mode of formation.

*Soil Survey Staff, Soil Conservation Service, USDA. *Soil Classification: A Comprehensive System,* 17th Approximation. Soil Cons. Serv. 2 and 24. 1960. Govt. Printing Office, Washington, D.C.

HORIZON

Most soils contain *horizons*. The horizon is a "layer of soil or soil material approximately parallel to the land surface and differing from adjacent, genetically-related layers in physical, chemical, and biological properties or characteristics such as color, structure, consistency, kind and number of organisms present, degree of acidity or alkalinity, etc."* All soils, though, do not contain discernible horizons. They vary in distinctness with surface topography and other soil-forming factors. Some boundaries are clear and sharp, but some grade into one another with no real differentiation or positive line of demarcation. Horizons vary in thickness. They do not necessarily relate to textural stratifications or bands as deposited by a transport agent or by parent rock formation, but to soil-formed three-dimensional layers smooth to highly irregular. Broadly speaking, the *O* horizon is the organic layer of mineral soils, usually absent in desert soils. The *A* horizon is a mineral layer in which organic matter accumulates and from which clay, iron, and aluminum move out, leaving an enrichment of quartz or other resistant minerals. The *B* horizon accumulates constituents from the *A* horizon; the *C* horizon is a mineral layer, excluding bedrock, considered to be the parent material for A and B. Very little effect of the soil-forming factors can be observed in C. The *R* layer is the underlying consolidated bedrock.

LIME

Desert soils contain lime (calcium carbonate). Lime may be detected by dropping acid onto the soil and observing the effervescence. The bubbles are carbon dioxide, like in soda pop, escaping from the carbonate part of the lime. Lime is dispersed throughout the subsoil in various forms as a powder, in soft or hard nodules, in veinings along root and worm channels, in tongues, and in soft or hard layers which are easy to see with the naked eye (Fig. 5.1). Lime also coats roots and soil particles, gravel, and stones, and cements them together. It tends to accumulate in a layer in the soil at a place near or at the mean depth of rain penetration. "Caliche"

*Soil Survey Staff, Soil Conservation Service, USDA. *Soil Classification: A Comprehensive System*, 17th Approximation. Soil Cons. Serv. 2 and 24. 1960. Govt. Printing Office, Washington, D.C.

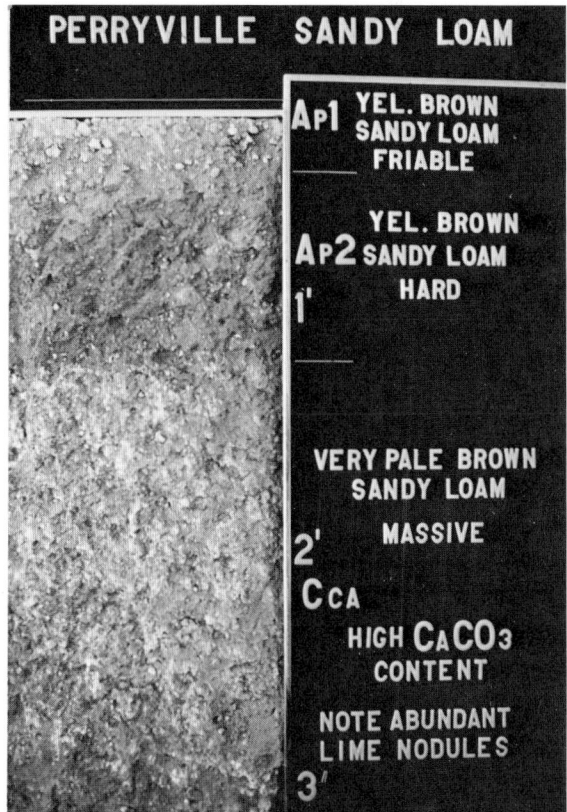

PERRYVILLE SANDY LOAM

AP1 YEL. BROWN SANDY LOAM FRIABLE

AP2 YEL. BROWN SANDY LOAM HARD

1'

VERY PALE BROWN SANDY LOAM

2' MASSIVE

CCA

HIGH $CaCO_3$ CONTENT

NOTE ABUNDANT LIME NODULES

3'

Fig. 5.1. Lime veinings and nodules in a Perryville sandy loam.

is the local name for limy layers, which range in thickness from a fraction of an inch to as much as 25 feet (Fig. 5.2).

The origin of massive, thick lime or caliche layers is believed to be fossil residue from a humid climate, accumulating from the calcium of geological material dissolved and redeposited on the bottom of lakes, playas, seas, and oceans. If the water has been inhabited by crustaceans or ostreans that accumulate crusts or shells of lime, these fossils will appear embedded in the caliche along with other lime-depositing living organisms. The Imperial Valley of the basin surrounding the Salton Sea provides an obvious example of a historic sea (ocean) bed. Soils formed over these deposits through the ages as water- and air-borne materials were deposited under different climatic conditions.

Arid and semiarid vegetation grow quite well in lime soils except where the layers are so near the surface and so hard that roots fail to penetrate, limiting the feeding volume. Domestic plants grow poorly, if at all, over shallow caliche. Soil moisture is the most limiting factor in shallow soils. In landscaping, holes must be extended into and

STRATIFIED
SANDY SOIL

LIME
LAYER

Fig. 5.2. Deep deposit of caliche below stratified alluvium.

through the caliche for good rooting and drainage. The holes should then be filled with suitable soil. Certain plants in limy soils require supplemental iron to prevent "lime-induced chlorosis," an iron deficiency symptom.

GYPSUM

Gypsum (calcium sulfate) is found in desert soils of the Southwest in small amounts, as crystalline granules precipitated alone or with lime; as veinings in root and worm channels (Fig. 5.3); or in great depths as relatively pure crystalline sand (Fig. 5.4). Gypsum dunes may cover large land areas, such as in the White Sands, New Mexico, and underlie wind- and water-deposited soils, as along the Pecos River in New Mexico and Texas. Gypsum is most prominent where parent rock contains sulfur or sulfate compounds. The origin of gypsum in soils is multiple. The precise origin of large gypsum sand dunes, however, still is not settled.

Gypsum is so insoluble in water, just as is lime, that its maximum solubility provides less soluble salt than is necessary to inhibit

Fig. 5.3. Gypsum veinings in soil root channels.

Fig. 5.4. Gypsum sand dunes (Ray Manley photo).

plant growth of arid land plants and urban landscaping plants. Technically, a saturated solution at 25 C contains 2,630 ppm of gypsum. Many plants grow well in a soil solution of this gypsum concentration. In fact, gypsum is added to sodium-affected soils as an amendment to replace the sodium with the more favorable calcium contained in its molecule.

Gypsum horizons in desert soils generally appear in specific, localized areas. Most irrigation water contains some gypsum which accumulates appreciably throughout the soil in time. Except where gypsum is shallow and forms a physical barrier, limiting sufficient root feeding zone for maximum plant growth, it is not detrimental to plants.

HARDPANS

Hardpan (duripan) horizons in desert soils offer serious inhibition to maximum plant growth when they occur within the feeding zone of roots. The word "pan" means a hard layer of varying thickness. The nearer the pans are to the surface, the greater the retarding effect on growth. Hardpans may be compacted clay layers, silica cementations, silica and lime cementations, lime (caliche) alone, or lime and/or gypsum indurations. These layers must be penetrated and broken up for most effective plant growth.

Caliche hardpans of the Southwestern deserts are thick, hard, and impenetrable to roots. Subsurface stones and rocks frequently imbed with caliche. These are unusually difficult to break up and eliminate so plants can grow.

Claypans

Silicates are an important feature of desert soils, and are seen as the highly insoluble silica dioxide and other oxides, silicate sand, and clay materials. Because of the alkalinity in some soils, silicates such as those of sodium and potassium become soluble and precipitate in the lower horizons in less alkaline lime (caliche) layers. Under arid and semiarid conditions silicates have been suggested as contributing to the problem of hardness and impenetrability of caliche. Small quantities of soluble silicates are usually found in water extracts from soils of the desert Southwest. The more alkaline the soil, the greater the amount of silica in solution.

Clay layers high in silicate clay and silica may be found in sub-surface horizons. They may originate by slow weathering in place under either arid or semiarid conditions but, for the most part, are remnants of a more rapid weathering of a humid climatic period. These horizons or layers often become compacted and inhibit water movement and leaching of salts. Clay soils as well as clay layers often resist effective root penetration and are poorly aerated, salty, and of low water permeability. Salts concentrate in clay layers because there is a greater amount of pore space. Water acts as a vehicle carrying salts to the clay as soils dry out. Claypans should be broken up and mixed with other, more coarse soil particles within the root-feeding zone for best plant growth.

Ironpans

The red and brown colors of desert soil originate from dehy-drated iron hydroxides and iron oxides of a specific hydration level. Iron comes from rock minerals which dehydrate and oxidize rapidly under the high temperature conditions of the desert after periodic wettings. Iron hydroxides appear as films or touches enveloping the soil particles and sands, and form scales and coatings on stones and cobbles. Soils of the Southwest contain a relatively high amount of "free" iron oxide. The total iron content often ranges up to 14 per-cent, calculated as Fe_2O_3. Iron hydroxides and oxides cement parti-cles together, and when concentrated form ironpans of even higher iron content. Roots penetrate ironpans with difficulty, if at all.

ALKALINITY

Soil reaction (or pH value)

Desert soils are near neutral to alkaline in reaction, i.e., they have a pH value of 7.0 to 10.0 (see Fig. 3.7). Most soils in the deserts of the Southwest range in pH values of about 6.8 to 8.4. Numerous plants grow favorably within these pH limits. Small, localized areas of sodic (sodium) or potassic (potash) soils have pH values above 8.4. Other salts may be found in excessive concen-trations in limited areas of desert land, but sodic soils require special attention for reclaiming. Except for leaching of excess salts with water, very little can be done economically to alter the pH much from its original value except in small flower beds and planters around the home where economics do not prevent additions of special soil amendments to lower the pH value. Acid-loving plants still must contend with salty irrigation water indigenous to the area and a hostile desert climate, even if the bed of soil in which they have their roots has been made favorable.

Sodium (Black alkali)

Sodium salts in particular create a problem in soils since they tend to disperse the particles, puddle them, and solubilize humus. Moreover, they impart a high pH value to soil where they accumulate. Sodium-affected soils (sodic or natric) are characterized by the presence of a layer having its cation-exchange positions occupied, 15 percent or more, by sodium. They have been referred to as *Solonetz* or *solidized-Solonetz* soils by early soil scientists, particularly the Russians, who studied them in eastern Europe and Asia. The sodium-affected subsoil layer has a prismatic, columnar, or block structure with rounded biscuit-like tops (Fig. 5.5). These soils must be reclaimed, using a calcium-yielding salt, such as gypsum, or certain amendments that release soluble calcium from lime, such as sulfur or sulfuric acid, as well as employing extensive leaching (washing) to replace the sodium by the calcium.

Excessive potassium salts cause black alkali conditions just as sodium does, although potassium soils occur less frequently. The peculiar subsoil of sodic and potassic soils is unmistakably different from those of any humid climate.

SALINITY, SALTY

The sparse rainfall of deserts results in a greater build-up of salts released from weathering rock in arid than in humid soils. Crusts and fluffs of white powdery salts scattered on soil surfaces depict saline conditions in the subsoil. Even where no salts appear on the surface or within the root zone, irrigation will tend to bring salts up to the root area by capillary action unless the water penetrates deeply. Often salts concentrate in layers where texture changes occur. Lack of adequate drainage and high water tables allow salts to move upward with consequent harmful effects on even the most tolerant plants. When salts collect in large enough amounts to seriously affect plant growth, soils are called saline or "white" alkali.

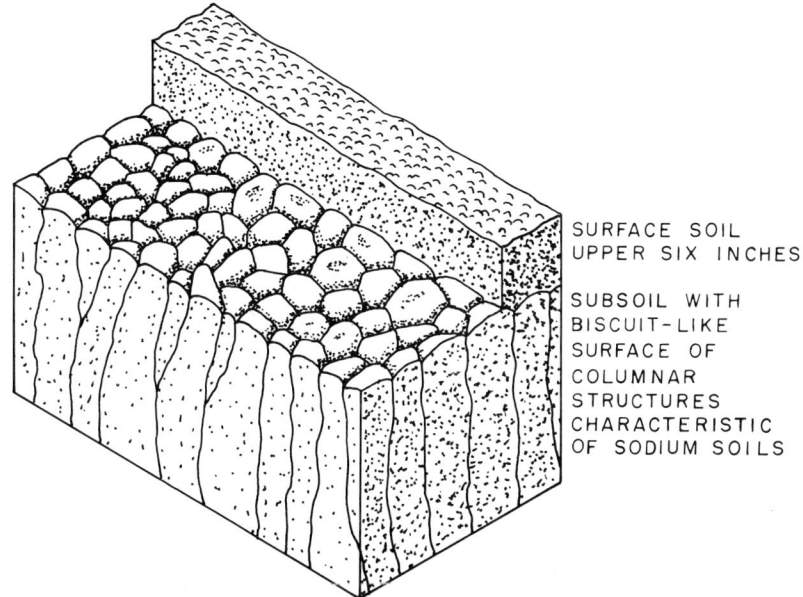

Fig. 5.5. Three-dimensional soil block showing tops of columnar structures in the subsoil where sodium dominates.

If the salt is predominantly sodium and the humus is solubilized, making the soil black, soils are called "black" alkali or sodic (Fig. 5.6). They must be reclaimed to be productive. The excessive accumulation of some salts such as those of sodium and potassium has a detrimental effect on the physical condition of the soil, making root penetration difficult or impossible.

Plants differ markedly in their ability to tolerate salt concentrations. Lists of salt-tolerant plants have been prepared for distribution at many of the western land grant universities.

STRATIFICATIONS

Texture: gravel, sand, silt, clay

Alluvial soils of the valley floors on which the cities of Brawley, California, Las Cruces, New Mexico, Phoenix and Tucson, Arizona, and others are located contain stratified sand, silts, and clay or gravel (Fig. 5.7). For example, a few inches of loam surface may overlie layers or islands predominating in silt, sand, or clay. Thin lenses of clay, even a fraction of an inch thick, can slow down the rate of water flow through the soil. Water will collect and perch wherever a pronounced textural change occurs, even though, for example, it be silt over sand or sand over silt. Abrupt textural changes such as these should be eliminated by deep tillage and mixing to make the soil more homogeneous within the root zone.

Although textural heterogeneity is common, alluvial soils of broad valleys may be quite uniform in texture and structure and relatively free of subsurface horizon conditions. These soils produce well and are ideal for residential and municipal landscaping without special mechanical manipulations.

Lime, clay, and gypsum combinations

Caliche, clay, and gypsum, stratified in thin and thick layers, appear in many soils either together or alone along drainageways, rivers, streams, and in alluvial fans (outwashes). Soils may vary quite radically in the distribution and thickness of these layers. Their extent in alluvial soils cannot be predicted with accuracy except by probing with a soil auger and observing the nature of cores taken from the profiles. Such distinct layered variations are identified in a

Fig. 5.6. "Black alkali" soil where sodium salts dominate. Note the biscuit-shaped tops of the columnar structures.

Fig. 5.7. Stratified alluvium along the Santa Cruz River south of Tucson, Arizona. Note man (arrow).

soil survey program, delineated, and located on a soils map of the area. These maps are available from land grant universities and the USDA Soil Conservation Service state offices.

PIPING

Piping is a "tunneling erosion" where subsoil loosens up and flows when the soil becomes saturated with water. Buildings and highways have been known to be threatened with foundation instability and loss by piping action. Piping causes sinking and caving of the soil in "tubular" or "pipelike" depressions (Fig. 5.8). Piping can be a serious problem to plantings too close to buildings, as well as to residential and urban buildings themselves.

Alluvium, where differential particle-size distribution and deposition have taken place and where subsoils are highly dispersed by certain salts, is susceptible to piping. Earth shrinkage and cracking due to removal of underground water over long periods of pumping may initiate piping formation, as wind and water deposit fine sandy material that differs in texture from the surrounding soil in the resulting cracks.

Fig. 5.8. An example of soil piping. Note the arrow pointing to the hole in the side of the ditch where water has forced the soil to move in the pipe and out the exposed end.

Because piping is due primarily to differential particle-size distribution in the peculiar pipe pattern, it may be controlled by any method that will mix the soil to a uniform, homogeneous mass. Where piping is suspected, deep plowing, ripping or tilling with blades or farm implements is suggested before building foundations are laid and landscapings begun. Control is easily accomplished on farmland.

SMALL MAMMALS, WORMS, INSECTS

A vertical cross section of a soil reveals the activity of small mammals, worms, and insects. Burrowing mammals form tunnels which fill again with soil from above and in time may greatly modify the profile by mixing the soil. Rodent channels frequently act as water channels, causing fields to lose water or flood in unwanted areas. The activity of burrowing worms and insects establishes drainage channels, which often aid plant rooting. Such worm and insect channels act as avenues for air, water, salts, and silt. Even in the desert the earthworms work and rework the soil when it is wetted by the infrequent rain. They and other worms, beetles, and ants bring a large amount of subsurface soil, sand, and fine gravel to the surface. They also drag plant residue, seeds, and debris deep down into the soil. Termites, which are unusually active in soils of the desert

Southwest, digest cactus, grass, and the twigs and bark of shrubs. Their mud coating on surface debris is particularly obvious after summer rains.

Insect action can also be detrimental to vegetation or crops. Figure 5.9 shows patches of barley stripped from the land by ants. As much as 20 percent of some fields in the past have been permanently denuded. Ants now can be kept under control with insecticides new to the market, except, of course, where effective insecticides have been prohibited.

YOUR SOIL

You, the homeowner, landscape architect, and turf expert, want a soil that has a deep root-feeding zone and good internal drainage and aeration, as well as freedom from hardpans, shallow caliche layers, stones, excessive salt, alkali, or accumulation of toxic elements, and a good supply and balance of nutrients. Fortunately, the soils of the desert Southwest are fertile and productive in the broad valley areas with only a limited number of inherent problems. Arid and semiarid uplands and narrow valleys, however, possess some indigenous soil problems such as shallowness, caliche, hardpans, etc.

Fig. 5.9. The bare spots are ant hills in alfalfa.

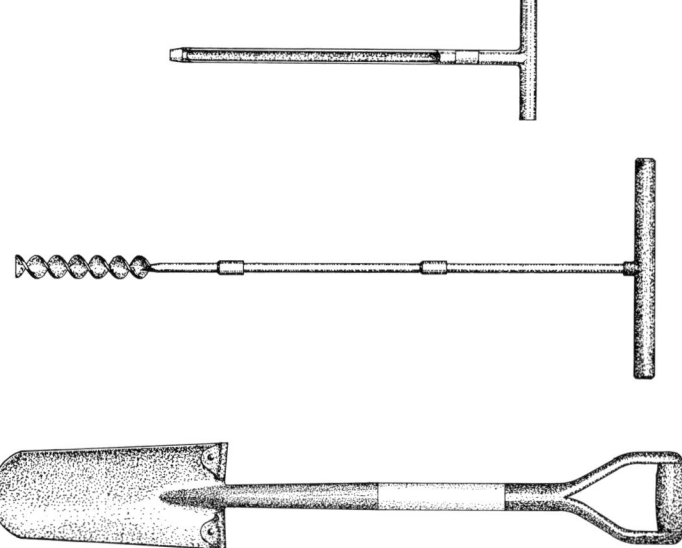

Fig. 5.10. A soil probe, auger, and spade used to obtain soil samples and probe the soil for subsoil characteristics which may require attention to provide suitable plant growth.

Plant-growing systems for these conditions can be developed by those knowledgeable in arid soils.

The most single limitation for plant growth in virgin soils of the desert Southwest, other than water, is the deficiency of nitrogen, which must be added to get the best growth.

Caliche layers must be dealt with to insure sufficient feeding area and drainage when deep-rooted trees and shrubs are planted.

Lime often accentuates iron chlorosis or yellowing of sensitive plants. This lack of plant-available iron can be readily corrected by adding iron sulfate or "chelate."

Knowing the characteristics of the soil below the surface (root-feeding area) is much more essential in desert plant-growing than in humid climates. Soil and water management practices, necessary for long-term productivity, require that the soil profile be characterized to a depth of at least three feet. A soil auger or probe (Fig. 5.10) can be helpful in obtaining this information, since it is not practical for the grower to dig pits except for tree and shrub plantings. Information about your soil, when obtained before permanent structures are placed and irrigation systems installed, will eliminate a lot of headaches later when more soil-corrective measures may be required or limitations imposed in landscaping choices.

6. Soil Classification

Not all of the earth's loose crust is considered soil. Some material may not be developed significantly from what it was when originally exposed as broken stone. In fact, certain desert areas have formed no layer which can be classified as soil at all, and the mineral debris occurs as expanses of shallow broken rock overlying bedrock, or as mineral particles which occur as shifting sand dunes. Thick beds of lime or gypsum expose areas which have no trace of soil development. Salt-encrusted spots, bare of vegetation, also may be found scattered on desert valley floors with little to no recognizable soil profile development. Soils form in time, however, from the earth's loose crust, through its interaction with the soil-forming factors of temperature, moisture, plant life, and topography. Land must support plant growth to be classed as soil.

Desert soils are difficult to group, because both arid and humid climatic conditions prevailed at one time or another in their history. Moreover, transport, mixing, and erosion processes, which persist to this day, scatter even the most well-developed soils beyond recognition. A soil well-developed under a humid climate, for example, appears in Figure 6.1 at a New Mexico desert site. It has been buried for millions of years below lava flows and ocean- or lake-deposited material. It provides evidence of a historic climate unlike that of today. Research shows that iron and aluminum clays and calcium have moved out of the A horizon and concentrated in the B, leaving the upper horizon higher in silica than the original geological material.

CLASSIFICATION AND DESCRIPTION

From information about surface and subsurface characteristics, soils of the desert Southwest are grouped into a national and indeed a world scheme of classification, just as are plants and birds. The subsurface horizon and stratified layers of different substances and textures provide essential characteristics for identifying purposes.

Aridisols. The name *Aridisols* is derived by combining the words *arid* and *soils*. They occur in dry places, have pale-colored surfaces which are normally soft when dry, and have no distinct structure. Aridisols must have one or more identifying horizon(s):

> *argillic:* silicate clay concentrations
>
> *natric:* sodium in the subsurface causing particle dispersion and silicate clay concentrations
>
> *cambic:* light color, low organic matter, fine texture with altered structure, not rock origin
>
> *calcic:* lime accumulation layer in the profile
>
> *salic:* salty, soil of salts more soluble than gypsum
>
> *duripan:* silica-cemented layer, hardpan or duripan
>
> *gypsic:* gypsum accumulation layer in the profile

Aridisols are subdivided into Argids and Orthids.

Argids are soils of dry places which have a clay (silicate clay) layer or horizon in the soil proper. In addition they may have a sodic, calcic, gypsic, salic, or duripan horizon.

Orthids in contrast are less well-developed than the Argids, lacking a clay (silicate clay) layer, although they may have other identifying layers.

Entisols (alluvial material). Most people recognize *Entisols* under names such as sand dunes, stream- and wind-deposited sandbanks, and river bottom or terrace material brought in or eroded by recent floods. There are no developed layers in Entisols that can be used for identification purposes. They resemble the parent material, from which they inherit profile properties. Entisols in the desert Southwest include three suborders: Psamment, Fluvent, and Orthent.

Psamments are air-transported sands such as dunes or sandy beaches sorted by water not possessing identifying layers.

Fluvents are water-transported and sorted materials (alluvials) which occur in flood plains, fans, and deltas of rivers and streams.

Orthents are residual soils which appear on land slopes of erosional surfaces (*Lithosols*). The loss of the surface leaves a shallow, eroded, and/or truncated soil. In arid and semiarid lands the soils are neutral to slightly alkaline and often limy.

Mollisols. The term Mollisol comes from (mollis) soft and soil. *Mollisols* are soft, dark-colored, high organic matter soils lo-

Present desert surface

Stratified old sea bed covered the area after volcanic action and soil formed

Wedge of volcanic rock deposited after soil formed

Buried soil profile formed during humid climate

Fig. 6.1. A buried soil profile formed in ancient history during a humid climatic cycle, New Mexico.

cated along river channels and in flood plains or terraces subject to floods. Only one Mollisol suborder, Ustoll, occurs in arid lands.

Ustolls possess a thick-dark organic surface layer with powdery lime. Sodium and other salts often accumulate because of their low position along drainageways and rivers.

ARIDISOLS

Soils of dry places are grouped under the heading of *Aridisols*. Commonly associated with aridisols, which have distinct identifying horizons or layers, are barren sands, salt flats, and playas. Figure 6.2 is a photograph of a clay playa, much of which is barren of vegetation.

Fig. 6.2. The Willcox Dry Lake, a clay playa which is both saline and alkaline and so high in clay that water fails to leach it (Ray Manley photo).

Wind influences aridisol development. Fortunately for residents in the desert Southwest, high velocity wind storms of dust and driving sand are less extreme than in other world deserts. Nevertheless, sand and dust move quite violently on occasion, leaving a desert pavement of polished pebbles, denuded rock and stone, moving sand dunes, or deposits of *loess* (wind-laid dust).

Rainfall is so scant in the desert Southwest that it leaches lime only to a shallow depth. In the Yuma and Imperial valley areas, for example, rainfall isn't sufficient to remove all the lime from the surface particles. Where lime has accumulated into a noticeable layer (either as lake or sea deposition, or by rain solution and reprecipitation in a lower layer), it is locally called "caliche" (see Figs. 5.2 and 5.3 in Chapter 5). Lime layers appear at various depths, distributed unevenly in the soil. Gypsum usually deposits similarly, either with lime or alone in a more scattered, granular condition. Deep gypsum layers and concentrations are found frequently along the Pecos River in New Mexico and Texas. Gypsum appears in many soils of the desert Southwest and is thinly scattered throughout the soil profile in association with other salts. Lime deposits, underlying most desert soils, vary greatly in thickness. Obviously a one- or two-foot soil layer does not give rise to three-, four- or ten-foot layers of lime. Differences in thickness of lime, the presence of salt deposits, fossils, clay lake bed formations, and historic pollen provide evidence that the desert was exposed from time to time to a more humid climate. In fact, buried, truncated, and degraded remnants of well-developed soil profiles with clay accumulations lead one to believe that wet and dry cycles were the rule. Sometimes two and three well-formed profiles can be seen following one another, each with a climatic history of many thousands of years. Stabilization of such land took place during humid glacial periods and interglacial periods of dry climate. Ancient soil relics have been less altered in deserts than in humid areas.

Silicate clay layers and other layers of possible humid-climate origin like duripan, hardpan, and limestone, add futher variations to desert soils. Recent deposits of water-washed debris from overflowing streams and arroyos during torrential rains can change soil characteristics in a very short time. Soils in alluvial fan (arroyo and stream outflow) positions may change continuously as they are covered, uncovered, and dissected by flash storms. Sparse vegetation exposes arid lands to all the erosive processes of nature, which give the desert a feeling of restlessness.

Aridisols are technically subdivided into two suborders, *Argids* and *Orthids*. Most soils that have been called desert soils in the past — Red Desert soils, Brown soils, Reddish Brown soils — are included in the Aridisol class.

Argids

The silicate clay layer in Argids, which developed as an accumulation within the subsurface, is referred to as the B or illuvial horizon. Figure 6.3 shows a soil of the Lehman series which has a shallow argillic layer over bedrock. Sometimes this layer becomes exposed by arid climate erosion. The great groups which are subdivisions of Argids include:
1. Those soils of clayey layer that do not have a sodic or hardpan layer (soil series Stellar loam and Mohave loam, Figs. 6.4 and 6.5.).

Fig. 6.3. This soil belongs to the Lehman Series, which is characterized by an Argillic or clay layer. It is a shallow soil with a clayey horizon over bedrock and lime.

Fig. 6.4. Stellar loam from New Mexico. It has no hardpan or sodic layer, but contains a clay or Argillic layer.

2. Those soils of clayey layer that have a hardpan but no sodic layer (Palos Verde gravelly sandy loam).
3. Those soils of clayey layer that have a sodic layer but no hardpan (Harqua stony loam).
4. Those soils of clayey layer that have a sodic layer above a hardpan.

Orthids

The Orthids have layers similar to Argids except they lack a pronounced B horizon of clay accumulations. In their place is a changed or altered horizon, called *Cambic*. The material in the Cambic horizon is altered enough so that it shows little evidence of the rock structure from which it originated. It forms a structure of its own if the texture is suitable. All of the layers or horizons described

Fig. 6.5. Mohave loam is one of the most agriculturally productive soils in the Southwest.

CAVE GRAVELLY LOAM

SHALLOW A1-C1 HORIZONS
OVERLYING WHITE GRAVELLY
STRONGLY CEMENTED CALICHE

Fig. 6.6. Cave gravelly loam
from Arizona. It has a hardpan
beginning at about the
2-foot depth.

appear in the subsurface. The Orthids include many of the soils originally grouped as Desert soils, Red Desert soils, and Sierozems in the United States. Orthids do not possess a sodic (sodium) or argillic (silicate clay) layer or horizon. They are more representative of the present environment than the Argids.

In brief, the layers or horizons present in Orthids, and some profile examples, are:
1. Water-soluble materials
    lime
    gypsum or salt (Reakor loam)
2. Petrocalcic horizons (cemented lime) softened by a single treatment of acid (Cave gravelly loam, Fig. 6.6)
3. Cambic, or horizon altered from its original characteristics

ENTISOLS (Alluvial material)

Entisols exhibit no natural distinctive horizons or layers which may be used for identifying purposes. In the desert Southwest, a small accumulation of secondary carbonates, sulfates, or soluble salts may have taken place in some Entisols. The amounts, though, are so small as not to constitute an identifying layer. Sand mesas, dunes, and undifferentiated sand alluvium along active streams exemplify typical Entisols. Alluvial fans and constantly moving dunes that may deposit, erode, and reappear represent the extreme of young Entisols. The most notable suborder of Entisols in arid climates is the *Psamment*.

Psamments

Psamments are well-sorted sands such as typical sand dunes (shifting or stabilized), cover-sands, or older sandy parent material which is primarily quartz that has not weathered appreciably (Fig. 6.7). Sandy beaches and levees separated or sorted by water action qualify as Psamments where no identifying layer or horizon has formed. These restless sands blow and drift readily, are droughty, and have low water-holding capacity. Psamments either drift into dunes or blow out as depressions. Not all sandy-textured Entisols are grouped with Psamments. Gravelly sands that do not blow readily are included in another group (Fig. 6.8). Psamments have less than 35 percent (by volume) of gravel or coarse fragments in all sub-horizons.

Fig. 6.7. Sand dunes just west of the Colorado River in California. A Psamment type soil material.

Fig. 6.8. Arizo gravelly sandy loam is a typical Entisol of gravelly sands which do not blow as readily as dune sand.

Fluvents

Just as Psamments are related to air transport of soil materials, *Fluvents* are related to water transport (alluvial) of soil materials. Fluvents form as brownish to reddish soils in recent water deposits; mostly in flood plains, fans, and deltas of rivers and streams, but not in swamps where drainage is poor. Fluvents may be flooded frequently if not protected. Their subsurface exhibits stratifications of differently textured materials, such as sands, silts, and clays, depending on the flood history (Fig. 6.9).

Fluvents of hot arid climates are described in the USDA Soil Conservation Manual of 7th Approximation Classification as: "— are not flooded frequently or for long periods. They have a torric moisture regime, and most are alkaline or calcareous and in places somewhat salty. The larger areas, having favorable topography and location close to water, are commonly irrigated. The natural vegetation in the United States consists mostly of xerophytic shrubs and cacti, but in some parts of the world the only vegetation on the soils has been irrigated crops because the sediments have accumulated while the soils were being cultivated." Arid Fluvents were called alluvial soils in the past.

Good examples of Fluvents are *Anthony* in Arizona and *Gila* in New Mexico (Fig. 6.10).

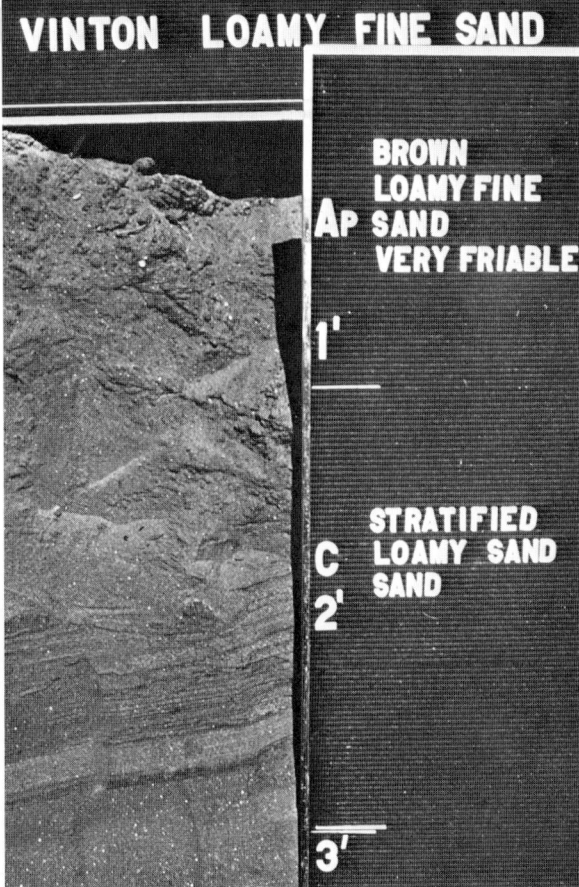

Fig. 6.9. Vinton loamy fine sand from New Mexico has been deposited by water and is highly stratified in texture.
A Fluvent soil.

Fig. 6.10. Gila loam from New Mexico, a Fluvent soil.

Orthents

Orthents occur on recent erosional surfaces. The loss of surface soil may be accelerated by cultivation or may be the result of geologic washing. Orthents of arid lands appear on moderate to strong slopes or rock pediments and are neutral to limy in reaction. Some are salty, but these occupy the more gentle slopes. Vegetation is sparse, consisting primarily of grasses, ephemeral forbs, and desert shrubs. Orthents (formerly known as Lithosols) contribute to the Southwestern economy by supporting a limited amount of vegetation for grazing. Local, state, and national lands control most of the areas

Fig. 6.11. An Orthent from California. It is a shallow residual soil developed in place over rock and lime.

occupied by Orthents, which are extensive in the desert Southwest. In contrast to Psamments, which are primarily *wind*-laid, and Fluvents, which are *water*-laid, Orthents form in place as *residual* soil; shallow, eroded and/or truncated. Millions of acres of rocky and rough land in the desert Southwest fall into this grouping. A few areas of Orthents may be found on rough and stony alluvial deposits.

An Orthent in California is the *Indio* series. *Superstition* appears in both California and Arizona. New Mexico has a series called *Yana*. Figure 6.11 illustrates an example of an Orthent.

MOLLISOLS

The *Mollisols* are dark-colored soils with a thick, dark surface layer, usually dominated by calcium and magnesium as exchangeable ions. The desert Southwest is not a characteristic climate for Mollisols, but a few appear along river channels and in river flood plains where soil has eroded into low terraces and water is more abundant either as runoff rain or frequent river flooding. These soils, formed under grass vegetation, accumulate organic matter which gives them the dark color. The soil textures predominantly are silt or clay loams. Lime concentrates at some place in the profile as soft nodules, veinings, and/or thin coatings on structures.

Ustolls

The suborder of Mollisols most frequently seen associated with arid lands is called *Ustolls*. Ustolls possess a thick, dark surface layer and a soft, powdery lime layer. Some occupy low spots or depressions, which lead to an accumulation of salts and often sodium. When sodium is present a natric or sodic layer develops with a weak or strong characteristic columnar structure.

The Hathaway series in Arizona and Portales in New Mexico are examples of Ustolls. They do not occur in arid land west of the Colorado River. Hathaway clay loam is a fertile and productive soil found along river channels, is high in organic matter, and finely textured.

The Pima series originally was classed with the Ustolls, but man's invasion of the area created too dry a habitat for such classification to continue. Pima soils since have become alluvial with Mollisol characteristics (Fig. 6.12).

When free of salts and sodium, these soils support excellent plant growth. Because of their low, flood plain position and fine texture they should be tested for alkali and salts before using them for landscaping purposes.

SOILS OF IMPERIAL VALLEY, CALIFORNIA

Photographs of soils in the Imperial Valley, California, reveal so little of the actual differences in soil characteristics that diagrams illustrating the soil profiles were prepared instead in order to accentuate this important group (Figs. 6.13 to 6.15). The most obvious variations are in the texture, which changes abruptly from sand to clay or clay to sand or to silt.

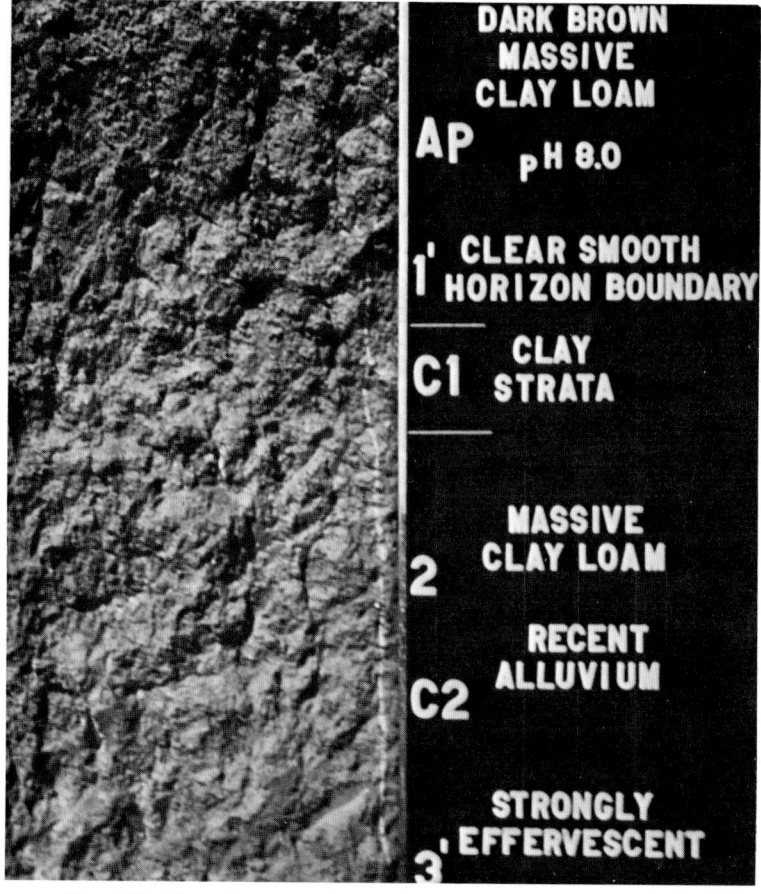

DARK BROWN
MASSIVE
CLAY LOAM

AP   pH 8.0

1' CLEAR SMOOTH
    HORIZON BOUNDARY

C1   CLAY
     STRATA

2   MASSIVE
     CLAY LOAM

C2   RECENT
     ALLUVIUM

3'   STRONGLY
     EFFERVESCENT

Fig. 6.12. Pima clay loam is an example of an Alluvium which appears along rivers, is clayey, and dark-colored because of a favorable content of organic matter.

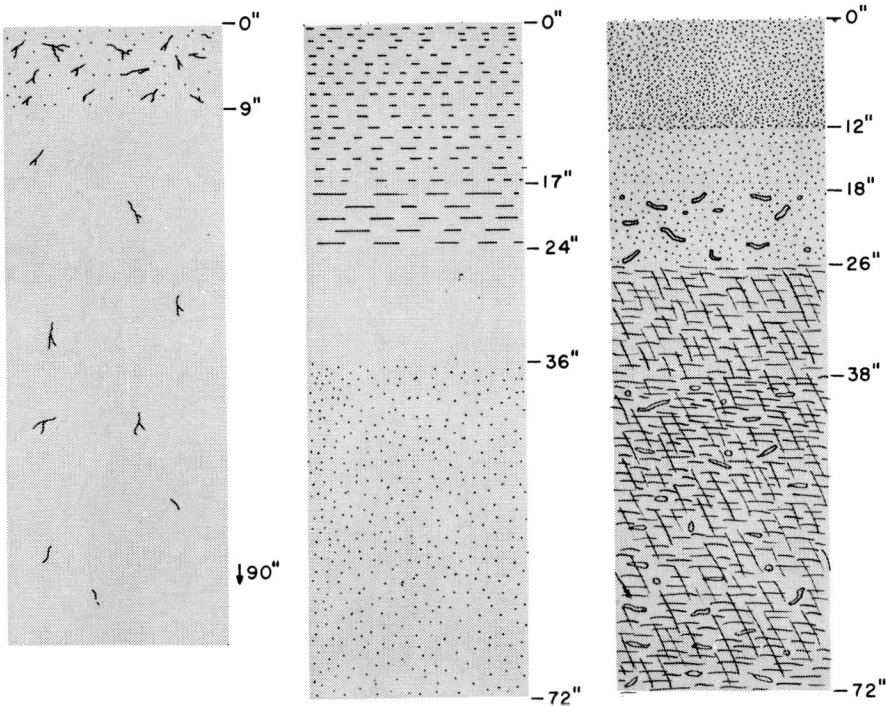

Fig. 6.13 (above, left). Rositas fine sand. The sands are reddish-yellow, strong brown when moist; single-grained, nonsticky; limy and moderately alkaline. From the surface to 9 inches deep, sand is very friable, and roots are common. Below 9 inches, sand is friable, few roots are seen, pH is 8.0 (many feet thick). The Rositas Series, developed on stabilized dune sands, are classified as Psamments, of mixed origin, having been deposited by wind and water. The type location is near the Imperial Irrigation District, Experimental Farm No. 2, Imperial Valley, California.

Fig. 6.14 (above, center). Holtville silty clay. These soils are clayey over loamy material which is calcareous and moderately alkaline. From the surface to 17 inches, the soil consists of light brown silty clay when dry, to brown and sticky when wet; massive structure; limy, moderately alkaline (pH 8.0); cultivated, hard. From 17 to 24 inches is light brown silty clay when dry, to brown and sticky when moist; thick platy structures; limy, moderately alkaline. From 24 to 36 inches is very pale brown silt loam when dry, to brown when moist; friable, slightly sticky; limy, moderately alkaline (pH 8.0). From 36 to 72 inches is very pale brown, loamy, very fine sand; massive structure, soft; very friable, nonsticky and nonplastic; limy, moderately alkaline (pH 8.0). The Holtville Series is located in Imperial Valley, California.

Fig. 6.15 (above, right). Meloland very fine sandy loam. The Meloland Series belongs to the Typic Torriorthents, coarse-loamy over clayey, mixed and calcareous. Meloland soils have a light color and are limy, moderately coarse-textured, with an upper horizon underlain by limy, silty clay within 35 inches of the surface. From the surface to 12 inches deep is light brown, very fine sandy loam, brown when moist; massive structure, very friable, slightly sticky; limy, mildly alkaline (pH 7.8). From 12 to 18 inches deep is very pale brown, loamy sand; pale brown when moist; soft, very friable; limy (pH 7.8). From 18 to 26 inches deep is very pale brown silt loam, pale brown when moist; massive structure; very fine rust-stained tubular pores; limy (pH 7.8). From 26 to 38 inches is pink silty clay, brown when moist; moderate, medium and thin platy structures; plastic; limy (pH 7.8). From 38 to 72 inches is pink silty clay, dark brown when wet; moderate, thick platy structure; hard, firm, very sticky; very fine tubular pores; gypsic and limy; mildly alkaline (several feet thick).

Appendix Section

APPENDIX A

Chemical Cycles and Soil Composition

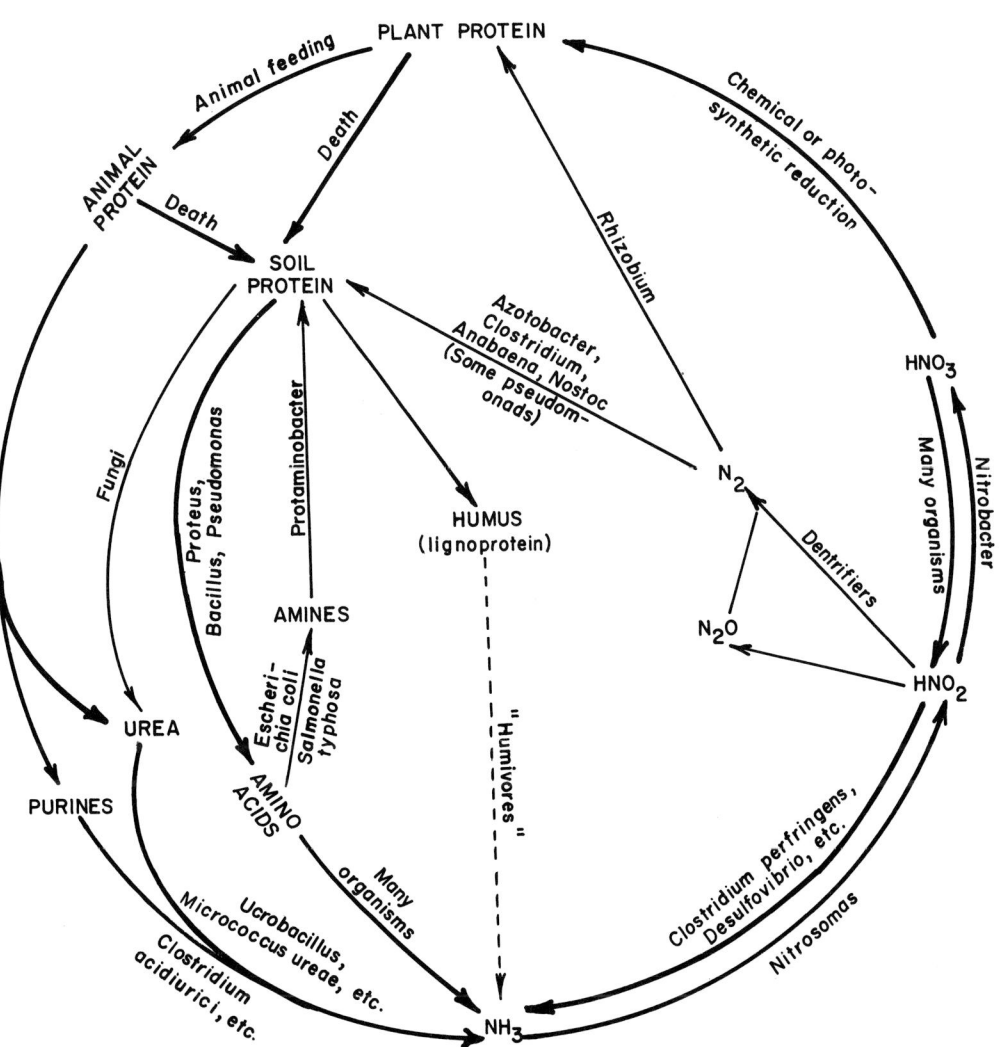

Fig. A-1. The nitrogen cycle in nature, showing microbial transformations. (Thimann, K.V. 1963. *The Life of Bacteria*. Ed. 2. Macmillan, New York)

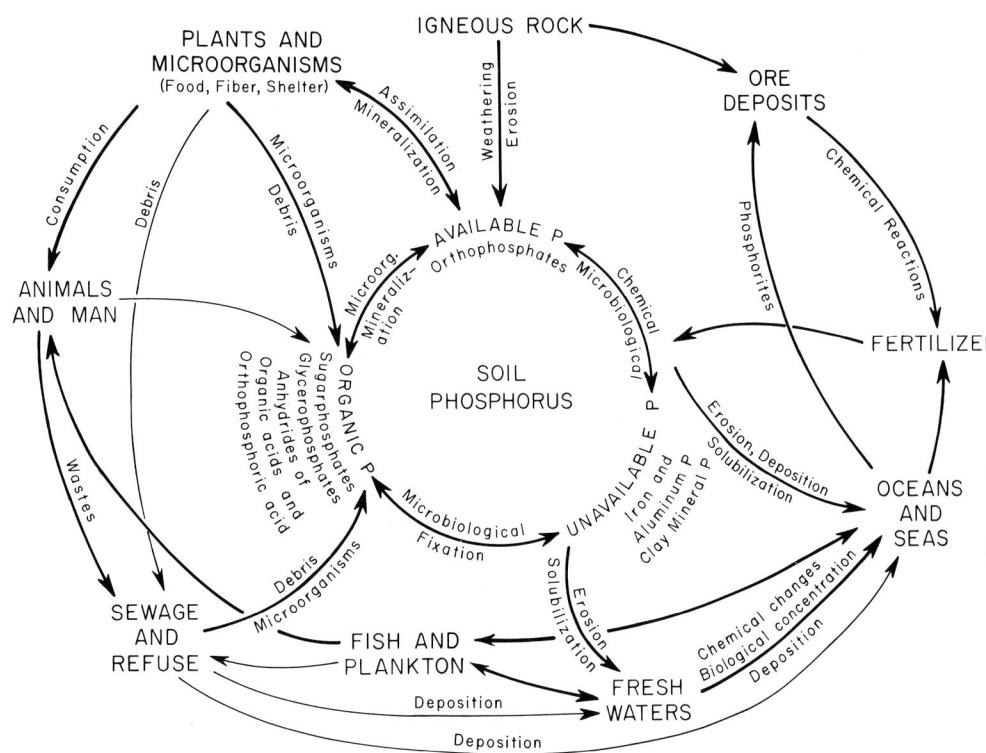

Fig. A-2. The phosphorus cycle in nature, showing the universality of phosphorus.

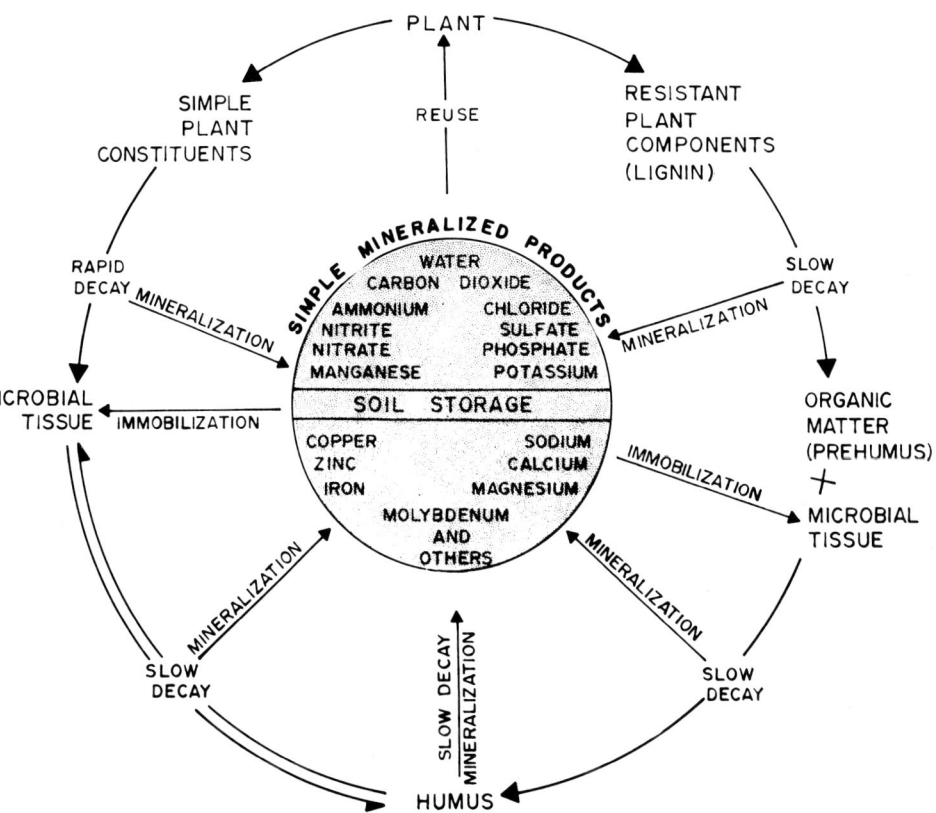

Fig. A-3. The carbon cycle in nature, showing decay, mineral nutrient release, and humus formation.

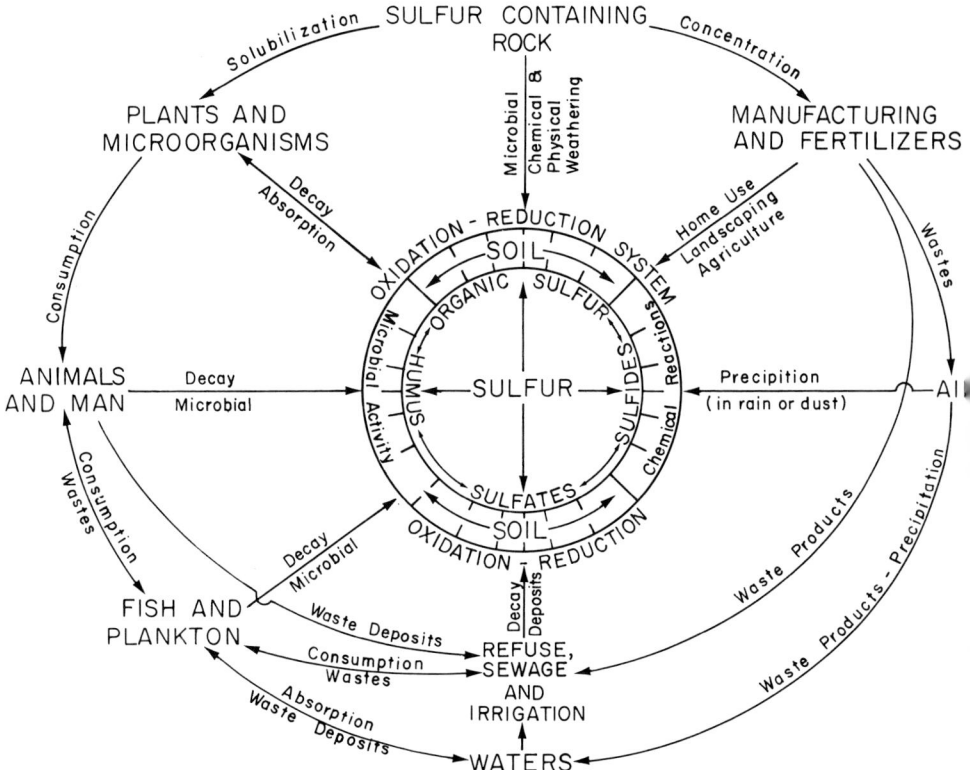

Fig. A-4. The sulfur cycle in nature, showing sulfur oxidation to sulfuric acid and sulfate, reduction to sulfides, and gypsum formation.

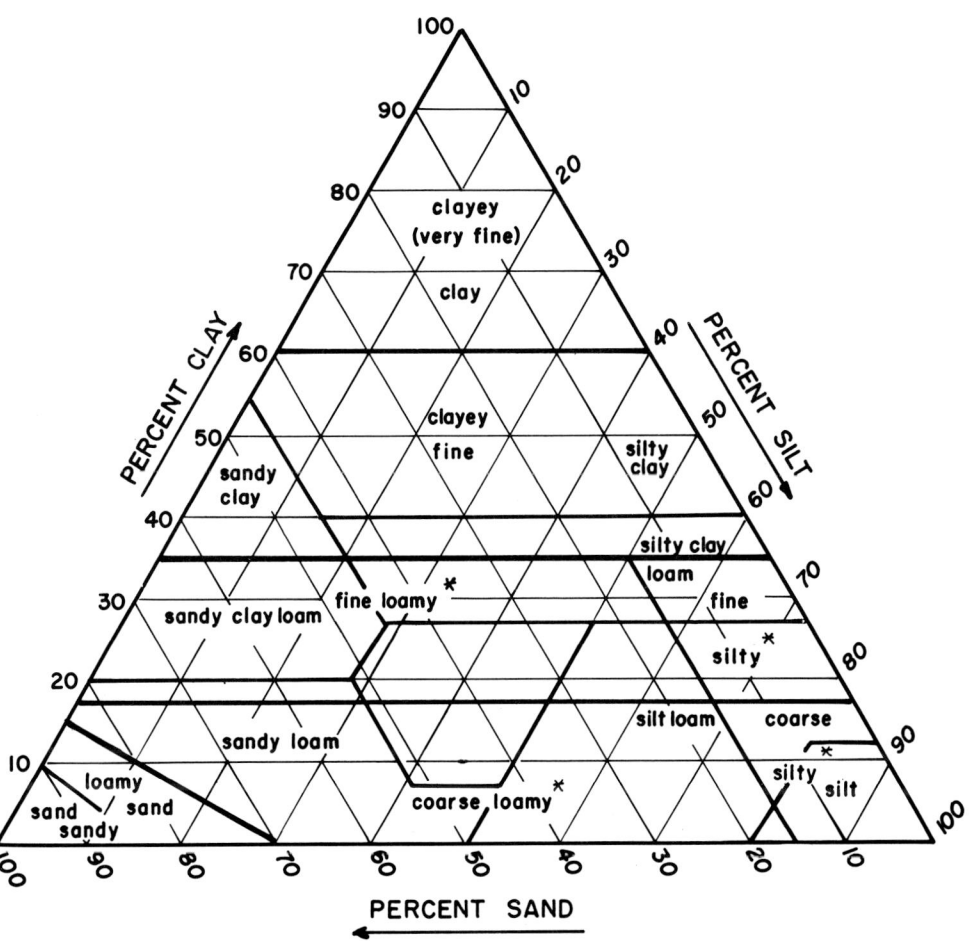

Fig. A-5. Texture chart showing percentages of particle sizes of soil separates and texture classes.

APPENDIX B

Conversion Factors for English and Metric Units

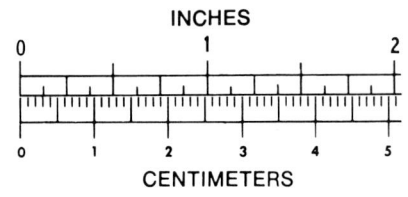

INCHES

CENTIMETERS

To convert column 1 into column 2, multiply by	Column 1	Column 2	To convert column 2 into column 1, multiply by
Length			
0.621	kilometer, km	mile, mi	1.609
1.094	meter, m	yard, yd	0.914
0.394	centimeter, cm	inch, in	2.54
Area			
0.386	kilometer^2, km^2	mile^2, mi^2	2.590
247.1	kilometer^2, km^2	acre, acre	0.00405
2.471	hectare, ha	acre, acre	0.405
Volume			
0.00973	meter^3, m^3	acre-inch	102.8
3.532	hectoliter, hl	cubic foot, ft^3	0.2832
2.838	hectoliter, hl	bushel, bu	0.352
0.0284	liter	bushel, bu	35.24
1.057	liter	quart (liquid), qt	0.946
Mass			
1.102	ton (metric)	ton (English)	0.9072
2.205	quintal, q	hundredweight, cwt (short)	0.454
2.205	kilogram, kg	pound, lb	0.454
0.035	gram, g	ounce (avdp), oz	28.35
Pressure			
14.50	bar	lb/inch^2, psi	0.06895
0.9869	bar	atmosphere,* atm	1.013
0.9678	kg (weight)/cm^2	atmosphere,* atm	1.033
14.22	kg (weight)/cm^2	lb/inch^2, psi	0.07031
14.70	atmosphere,* atm	lb/inch^2, psi	0.06805
Yield or Rate			
0.446	ton (metric)/hectare	ton (English)/acre	2.240
0.892	kg/ha	lb/acre	1.12
0.892	quintal/hectare	hundredweight/acre	1.12
1.15	hectoliter/ha, hl/ha	bu/acre	0.87
Temperature			
	Celsius	*Fahrenheit*	
$\left(\dfrac{9}{5}\,°C\right)+32$	−17.8C	0F	$\dfrac{5}{9}\,(°F-32)$
	0C	32F	
	20C	68F	
	100C	212F	

*The size of an "atmosphere" may be specified in either metric or English units.

Index

[99]